VOLUME FORTY SIX

INTERNATIONAL REVIEW OF RESEARCH IN DEVELOPMENTAL DISABILITIES

VOLUME FORTY SIX

International Review of RESEARCH IN DEVELOPMENTAL DISABILITIES

Current Issues in the Education of Students with Visual Impairments

Edited by

DEBORAH D. HATTON
Department of Special Education, Vanderbilt University, Nashville, Tennessee, USA

ELSEVIER

AMSTERDAM • BOSTON • HEIDELBERG • LONDON
NEW YORK • OXFORD • PARIS • SAN DIEGO
SAN FRANCISCO • SINGAPORE • SYDNEY • TOKYO
Academic Press is an imprint of Elsevier

Academic Press is an imprint of Elsevier
225 Wyman Street, Waltham, MA 02451, USA
525 B Street, Suite 1800, San Diego, CA 92101-4495, USA
Radarweg 29, PO Box 211, 1000 AE Amsterdam, The Netherlands
The Boulevard, Langford Lane, Kidlington, Oxford, OX5 1GB, UK
32 Jamestown Road, London NW1 7BY, UK

First edition 2014

ISBN: 978-0-12-420039-5
ISSN: 2211-6095

CONTENTS

SERIES PREFACE

For those who follow developments in this series itself, among the most interesting involves changes in the series' name. Starting with Deborah Fidler's topical Volume 40, we have changed the name of this series from the *International Review of Research in Mental Retardation*—the original name, in place from 1966 until 2011—to the *International Review of Research in Developmental Disabilities* or *IRRDD*. This name change, which some would argue was long overdue, reflects the current political realities and usages within the disabilities world.

But with this new name has also come related changes. As I noted in the Series Preface to Volume 40, the change from the term "mental retardation" to "developmental disabilities" opens up new possibilities. Developmental disabilities encompasses a wider terrain, now including the newest research advances in motor, intellectual, emotional–behavioral, or other conditions. If a condition begins during the childhood years (and hence is "developmental"), it is now within the purview of the *International Review*. As much as possible, I promised to reflect that larger, more expansive coverage within future *IRRDD* volumes.

The time has now come to make good on that promise. The current volume, Volume 46, is the first *IRRDD* volume devoted to Visual Impairments and Blindness. Under the steady editorial hand of Dr. Deborah Hatton, the volume features contributions reviewing the research concerning a variety of issues, established best practice interventions, and controversies related to educating children with visual impairments.

For those less familiar with visual impairments, a few preliminaries are in order. First, visual impairments and blindness are fairly rare during childhood. Unlike during the old-age years—when many experience a loss or diminution of vision—VI is considered a "low incidence" disability during childhood. Approximately 1 in 1000 school-aged children have visual impairments, and even smaller numbers have blindness.

A second, related issue concerns technology and the general needs of these children. Simply stated, the subject matter of visual impairments is specialized and technical. Teachers of students with visual impairments need to know about specialized computer software; braille (and changes occurring in different braille systems); various reading, math, and communication curricula and interventions; service animals; and orientation and mobility. Such knowledge is incredibly specialized, different even from the training special educators receive when they work with students with severe intellectual disabilities, autism, motor impairments, or multiple impairments. Moreover, given the low prevalence of visual impairments, few masters- or doctoral-level training programs exist and the professional community in visual impairments, while passionate in its dedication to these students, is very small.

Given this context, Dr. Deborah Hatton has devoted *IRRDD*'s Volume 46 to "Educational Issues for Students with Visual Impairments." Deborah has recruited an accomplished, research-oriented cast of experts to write on specialized topics to which each is knowledgeable, and then consulted with these professionals to complete their contributions. Once contributions were received, Deborah has, in her own gentle but firm manner, edited and helped to bring to completion the fruits of each group's labors.

Speaking for myself, for other non-VI *IRRDD* readers, and for the professionals and families who work every day with students with visual impairments, I thank Deborah for her hard work and dedication. As Deborah notes in her own Preface, the VI field has long been in need of a volume explicitly devoted to educational issues. But those of us outside of VI are also indebted, as Deborah has provided a glimpse into the world of visual impairments and the many educational advances that are currently benefiting these students. Enjoy this special issue on Educational Issues for Students with Visual Impairments, and, in a more general sense, welcome the long-running *IRRDD* series to research reviews on many different types of developmental disabilities.

Robert M. Hodapp
March 24, 2014

PREFACE TO SPECIAL ISSUE ON VISUAL IMPAIRMENT

Although formal educational programs for individuals with visual impairment have existed since 1784, the field has relatively little research to guide practice. In the United States, the few personnel preparation programs that prepare teachers of students with visual impairment often have one or two faculty members. Often, these faculty members have little time and support for research, accounting for the limited empirical base for educational practices. Syntheses of intervention research, and particularly evidence-based reviews, are of value in identifying practices that promote optimal student outcomes. Such syntheses can also help determine priorities for intervention research.

Some progress has been made on this front in the past decade. The chapters in this volume extend previous syntheses and demonstrate a variety of approaches to synthesizing research.

Ultimately, my primary goal for this volume is to provide an update on intervention research to identify effective educational practices for students with visual impairment. A second priority is to secure information to facilitate the current transition to the Unified English Braille Code (UEB).

In Chapter 1, I describe evidence-based practices for students with visual impairment within the context of the special education evidence-based practice movement. By briefly reviewing the history of education of students with visual impairment and initiatives that have shaped service delivery, readers gain insight into current opportunities for advancing the use of evidence-based practices. This chapter also presents recommendations for conducting high-quality research syntheses and for identifying evidence-based practices. Potential evidence-based practices for students with visual impairment are also discussed.

It is important to appreciate, however, that students with visual impairment are very heterogeneous, and this heterogeneity requires a range of interventions to address the varying needs of individuals with this developmental disability. Developmental outcomes have been shown to vary based on the presence of co-occurring disabilities, the amount of functional vision, and whether blindness was present at birth or was acquired at a later age. Approximately 60–65% of young children referred for specialized services have additional co-occurring disabilities; consequently, intellectual abilities can range from gifted to severe-profound disabilities. These wide-ranging abilities require educators of students with visual impairment to select and implement a wide range of interventions. For example, literacy goals for students with severe intellectual disabilities might involve communication interventions such as those described by Parker and Ivy in Chapter 4. Other students with average to gifted intellectual abilities might benefit from braille reading comprehension strategies such as those described by Savaiano and her colleagues in Chapter 6.

Development and learning is typically facilitated by visual experiences. Consequently, a top priority for students with visual impairment is to make certain that they use any available vision as efficiently as possible. The field's approach to optimizing available vision has changed drastically in the past 60 years and warrants a brief discussion. A notable development in the education of students with visual impairment occurred in the 1960s when Natalie Barraga showed that residual vision could not be "used up." Due to her advocacy, educators were persuaded to stop using *sight saving* approaches in which children were encouraged to "rest their eyes" so that they would not lose the little vision they had. In addition, Barraga developed strategies to help students use their limited vision as efficiently as possible.

Anne Corn, Barraga's colleague at the University of Texas for many years, extended our understanding of low vision and visual efficiency and the need for clinical low vision evaluations and optical devices for students. Corn's research demonstrated that some individuals can read regular print with optical devices (magnifiers) as efficiently as they read large print; she was among the first professionals in the field to question the use of large print with individuals with low vision. Her project, Providing Access to the Visual Environment (PAVE), provided a model of clinical low vision evaluations and services that included optical devices and support from teachers of students with visual impairment. Thus, Corn's work greatly enhanced access to the general curriculum and provided strategies for optimizing outcomes for students with low vision.

Although optical devices have been mentioned as a possible promising practice for students with low vision, access to optical and nonoptical devices to enhance vision seems like a basic right to ensure equal access to educational opportunities. Research suggests that individual preferences guide the selection of optical and nonoptical devices to facilitate reading, travel, and participation at school, home, and in the community. These devices and strategies to enhance visual function are one type of assistive technology for individuals with visual impairment. Assistive technologies to optimize vision can range from video magnifiers that employ the same technology as closed circuit televisions for both near and distance viewing to hand-held dome magnifiers and bold-lined paper and dark markers.

In Chapter 2, Smith and Kelly describe the current state of the literature on assistive technology to support classroom-based educational interventions. Assistive technology has made information much more readily available to individuals with blindness and low vision. Individuals who are tactile learners can now use screen readers and refreshable electronic braille displays to access digital documents, text on computer screens and personal electronic devices, and the Internet. By using accessible documents, e-mail, flash drives, and other mainstream technology, individuals who cannot read print can have ready access to the same information as their sighted peers. Expensive, specialized technologies for individuals with visual impairment, such as braille note takers and screen reading software, are being abandoned in favor of off-the-shelf devices that adhere to the principles of universal design for learning. Indeed, Apple's Voiceover feature for IOS devices, along with the ability to connect to refreshable braille displays, appears to be revolutionizing life for technology-savvy consumers who happen to be blind. Although assistive technology has clearly increased independence and employability of individuals with visual impairment, we have virtually no intervention research to guide our practice. Indeed, Smith and Kelly discuss survey research in which teachers of students with visual impairment repeatedly report that they do not feel prepared to teach assistive technology to their students. In addition, Smith and Kelly provide recommendations to enhance instruction in the use of assistive technology and to advance research on this topic. Clearly, rigorous research on assistive technology interventions is needed to inform educational practices and professional development.

For students with minimal useful vision, educators must use strategies based on tactile, auditory, kinesthetic, and other senses to help students with visual impairment achieve the same academic standards and similar levels of independence as their sighted peers. In Chapter 3, Ivy and I systematically review quality indicators of single subject research studies on systematic prompting with tactile learners. Researchers used prompting to teach a variety of skills to tactile learners with varying abilities. We conclude that systematic prompting is an evidence-based practice for students who are tactile learners. In the process, we provide a model for rigorously evaluating research articles using quality indicators.

Increasingly, communication is being viewed as the foundation for literacy. And, because literacy is the foundation for most content knowledge in academic subjects, Parker and Ivy's synthesis of research on interventions to facilitate early communication development is particularly important. In addition, because approximately 65% of students with visual impairment may have multiple disabilities, the information provided in Chapter 4 is relevant for students with a wide range of abilities, including those with deafblindness. Parker and Ivy synthesize information from 34 intervention research studies that educators can use to promote early communication development in their students with visual impairment.

It is also important to appreciate the changes occurring around the world in the field of visual impairment. Specifically, the United States is transitioning to the UEB; therefore, educators will face a number of issues related to braille during the next few years. The target date for the transition is January 4, 2016. Among English speaking countries that have adopted UEB, only the United States will use a second code for math, the Nemeth code. The Nemeth Code for Mathematics has been used in the United States for the past 60 years. There is literally no empirical research comparing the efficacy of Nemeth and UEB for math instruction. Therefore, educators face an ethical dilemma; should we introduce a separate code for mathematics when other English speaking countries around the world are using one unified code for literary materials and mathematics? The field needs high-quality research that documents that Nemeth code is superior and/or more efficient than UEB for mathematics to justify the use of two braille codes.

In Chapter 5, D'Andrea, Wormsley, and Savaiano review key literature on the implementation of UEB from other English speaking countries to inform the current transition in the United States. Many educators and administrators are anxious about this transition; however, the encouraging reports of implementation in other countries should allay anxiety.

In Chapter 6, Savaiano, Compton, and I document the limited research that has been published on braille reading comprehension. We discuss 12 studies that included measures of braille reading comprehension in the context of an empirical model of reading comprehension.

Educators of students with visual impairment have two primary roles: to ensure that their students can access the general curriculum through collaboration with general education teachers and families and to provide direct instruction in specialized skills such as braille, orientation and mobility (O&M), and the use of assistive technology to gain access to academic content and to prepare for college and careers. For the past two decades, the specialized content that is unique to students with visual impairment has been referred to by many educators within the field as the *expanded core curriculum* for students with visual impairment.

In 1996, Phil Hatlen introduced the term *expanded core curriculum* to describe skills that students with visual impairment must acquire in addition to the general education core curriculum. In Chapter 7, Lewis, Savaiano, Blankenship, and Greeley-Bennett provide a brief overview of the nine areas included in the expanded core curriculum. In addition, this chapter synthesizes intervention research in three of the areas: independent living skills, self-determination, and social skills.

Although O&M emerged as a rehabilitation service for veterans of World War II, most O&M specialists serve students in educational settings. However, limited intervention research on O&M has primarily focused on adults, not children. In Chapter 8, Wall Emerson and McCarthy identify current issues that impact O&M instruction. Specifically, changes in intervention arise from such issues as global positioning systems, quiet cars, and complex intersections. In addition to reviewing and synthesizing research on O&M, Wall Emerson and McCarthy provide recommendations for advancing O&M research and instructional practices.

Recognizing the need for more and better prepared faculty members in the field of visual impairment and sensory disabilities, since 2005 the U.S. Department of Education's Office of Special Education Programs has funded doctoral level preparation programs to increase the number of faculty. These doctoral students have received instruction in the characteristics of high-quality research and have been encouraged to conduct intervention research. Systems change initiatives, discussed in Chapter 1, combined with an increase in new faculty with preparation in conducting high-quality research, may result in more rigorous intervention research that can lead to effective practices. I hope that the findings from the research syntheses described in this volume may provide inspiration for lines of research for these new faculty members that will lead to effective practices for students with visual impairment.

Deborah D. Hatton
March 22, 2014

CONTRIBUTORS

Karen Blankenship
Department of Special Education, Vanderbilt University, Nashville, Tennessee, USA

Donald L. Compton
Department of Special Education, Vanderbilt University, Nashville, Tennessee, USA

Frances Mary D'Andrea
Braille Authority of North America, Pittsburgh, Pennsylvania, United States

Robert Wall Emerson
Department of Blindness and Low Vision Studies, Western Michigan University, Kalamazoo, Michigan, USA

Kitty Greeley-Bennett
School of Teacher Education, Florida State University, Tallahassee, Florida, United States

Deborah D. Hatton
Department of Special Education, Vanderbilt University, Nashville, Tennessee, USA

Sarah E. Ivy
Department of Special Education, Vanderbilt University, Nashville, Tennessee, USA

Stacy M. Kelly
Department of Special and Early Education, Northern Illinois University, DeKalb, Illinois, USA

Sandra Lewis
School of Teacher Education, Florida State University, Tallahassee, Florida, United States

Tessa McCarthy
Department of Education, North Carolina Central University, Durham, North Carolina, USA

Amy T. Parker
Western Oregon University, Monmouth, Oregon, USA

Mackenzie E. Savaiano
Department of Special Education, Vanderbilt University, Nashville, Tennessee, USA

Derrick W. Smith
Department of Education, University of Alabama in Huntsville, Huntsville, Alabama, USA

Diane P. Wormsley
Department of Curriculum and Instruction, North Carolina Central University, Durham, North Carolina, United States

CHAPTER ONE

Advancing the Education of Students with Visual Impairments Through Evidence-Based Practices

Deborah D. Hatton[1]
Department of Special Education, Vanderbilt University, Nashville, Tennessee, USA
[1]Corresponding author: e-mail address: Deborah.Hatton@vanderbilt.edu

Contents

Abstract

Increasingly, educators are being challenged to use educational strategies that have been proven to be effective through high-quality research. As could be expected of a low-incidence disability in a field with few researchers, educators of students with visual impairments have limited research-based information to guide practice. The purpose of this chapter is to selectively examine the current state of evidence-based practices for the education of students with visual impairments and to briefly review the context for the evidence-based practice movement. In addition to identifying evidence-based practices and potential evidence-based practices for students with visual impairments, recommendations for advancing the identification of effective practices for students with visual impairments are presented.

International Review of Research in Developmental Disabilities, Volume 46
ISSN 2211-6095
http://dx.doi.org/10.1016/B978-0-12-420039-5.00001-0

1. INTRODUCTION

The adage that "Nothing is certain except change" seems especially true for educators in the United States who are in the midst of multiple systemic change initiatives. These initiatives include the adoption of Common Core State Standards and high-stakes assessments to measure the attainment of standards in 45 states, the evidence-based practice movement in special education, linking teacher pay/evaluations to students' high-stakes assessment scores, and the move to more field-/clinical-based preparation of future teachers. Concern that students from the United States are not adequately prepared for college and career contributed to the Common Core State Standards Initiative of the Council of Chief State School Officers (2012). The evidence-based practice movement, which started in the field of medicine, was reflected in federal mandates (Individuals with Disabilities Education Improvement Act of 2004 and No Child Left Behind Act, 2001) for both general education and special education teachers to use "scientifically based practices." Evaluations that directly link student outcomes to teacher performance, or value-added assessments (Misco, 2008), have increased the anxiety levels of practicing teachers, future teachers, and university faculty members. In addition, changes in accreditation standards by the Council for the Accreditation of Educator Preparation (CAEP Commission on Standards and Performance Reporting, 2013) will require colleges and universities to provide more extensive field experiences. In addition, future teachers and the universities that prepare them will be evaluated in a manner similar to value-added assessments (CAEP Commission on Standards and Performance Reporting, 2013).

These changes in the field of education present a particular challenge for educators serving low-incidence populations. Nevertheless, these systemic changes may motivate educators to evaluate their practices and to adopt practices that have been shown to be effective through research. For that reason, the purpose of this chapter is to selectively, not exhaustively, assess the status of evidence-based practices and potential evidence-based practices for the education of students with visual impairments. To provide the context for the use of evidence-based practices, I will briefly discuss the education of students with visual impairments in the United States from a historical perspective.

2. EDUCATION OF STUDENTS WITH VISUAL IMPAIRMENTS

2.1. Eligibility for Services for Students with Visual Impairments

One of the eligibility categories listed in IDEIA (2004) is the term visual impairments, an overarching term that includes students with low vision and those with blindness. Eligibility for services in most states is typically determined by information from medical and functional vision reports that document visual acuities of 20/70 or worse in the better eye with correction or a visual field restriction of 20° or less. Although individuals within the field have used the terms *low vision and blindness* for the past few decades to describe subcategories of individuals with visual impairments, terminology can be confusing.

Anne Corn, a specialist in the area of low vision, used a definition of low vision adapted from the World Health Organization that seems to best capture the term for educators; individuals with low vision, even with the best correction, have difficulty in completing visual tasks but can enhance their ability to complete visual tasks with accommodations such as optical devices and environmental modifications (Corn, DePriest, & Erin, 2000; Corn & Koenig, 1996; Corn & Lusk, 2010). However, some states use the term low vision to refer to students whose visual acuity is between 20/70 and 20/200 (Huebner, 2000). The 20/200 criterion is related to the term *legal blindness*, another somewhat confusing term. In the United States, legal blindness is described as visual acuity of 20/200 in the better eye with correction or a field restriction of 20° or less. Individuals who meet either of the criteria for legal blindness may be eligible for resources from the American Printing House for the Blind through their schools or from other federal programs (Huebner, 2000). Some students with legal blindness have no useful vision and may be braille readers, others may be able to read regular print with optical devices, and others may be nonreaders due to severe multiple disabilities. Clarification and consistency of terms would be helpful for consumers, educational and medical professionals, and families.

2.2. Models for Providing Services to Students with Visual Impairments

Special residential schools for students who are blind were established in the United States in the early to mid-nineteenth century and probably can be said to represent the first formal special education services offered in the United States (Hatlen, 2000). Public school programs that included students with visual impairments were available as early as 1900 (Hatlen, 2000). The deinstitutionalization movement of the late twentieth century required students to be educated in the least restrictive environment, often interpreted as their local public schools, as described in Public Law 94-142 of 1975 (Education for All Handicapped Children Act, 1975).

Although the need for a continuum of educational placements was acknowledged in subsequent special education laws, the right to be educated in local schools with typical peers that was codified in 1975 changed educational service delivery models for most students with visual impairments. Local school systems, often encouraged by families and federal financial support, increasingly hired teachers of students with visual impairments (TVIs) who used an itinerant model. This increase in itinerant TVIs changed specialized schools for students with visual impairments. Some special schools went from serving 800 students per year to serving 100 or fewer. Consequently, many specialized schools became outreach centers to provide support to all students with visual impairments within states (Ajuwon & Oyinlade, 2008; Hatlen, 2000). Most specialized schools continue to provide direct services to students, and many local school districts with close proximity provide transportation for some of their students to attend such schools.

Nevertheless, since the late twentieth century, most students with visual impairments have been educated in local public schools, with special education services provided by itinerant TVIs (Ajuwon & Oyinlade, 2008; Hatlen, 2000). Large metropolitan school systems often provide resource rooms for students who need intensive braille instruction or specialized support that may not be provided as efficiently within inclusive classrooms (Koenig & Holbrook, 2000). Often, students with visual impairments are so attuned to auditory stimuli that they may have difficulty focusing on instruction from TVIs within general education classrooms. Also, the loud sounds produced by most braillewriters and braille embossers may distract students in general education classrooms.

The Elementary and Secondary Education Act (ESEA, 2002) required schools to demonstrate adequate yearly progress for all students, including those with disabilities (Harr-Robins et al., 2012). The ESEA resulted in improved scores by most children with special needs on benchmark assessments (Chudowsky, Chudowsky, & Keber, 2009). Legally, schools could no longer have low expectations for their students with disabilities. As noted, the current teacher evaluation reform makes the stakes even higher; value-added assessments will link individual teacher performance to students' standardized assessment scores (Koretz, 2008). Never has there been more motivation for teachers to use strategies that are evidence-based.

3. ADVANCING EVIDENCE-BASED PRACTICES

3.1. Special Education

To address the federal mandate for the use of practices shown to be effective through high-quality research, the field of special education launched a number of initiatives (Spooner, Knight, Browder, & Smith, 2012). Initially, the Institute of Education Sciences recognized only randomized controlled trials (RCTs) as sufficient to identify evidence-based practices in education. In response, leading special education researchers from the Division for Research, a subdivision of the Council for Exceptional Children (CEC), paid attention to the challenges of conducting high-quality RCTs with children with disabilities. Specifically, Odom et al. (2005) discussed the challenge of identifying evidence-based practices (EBPs) for children with disabilities. Concurrently, the Division for Research of CEC published a series of papers on quality indicators for research. Horner et al. (2005) described quality indicators for single-subject research, and Gersten et al. (2005) identified quality indicators for group and quasi-experimental designs; these publications have been of primary importance in identifying EBPs for the field of special education.

3.1.1 Systematic Process for Identifying EBPs

Along with the establishment of quality indicators for research, the field of special education recognized that having a few studies that support a particular practice was not sufficient for identification of EBP (Cook & Cook, 2013; Gersten et al., 2005; Horner et al., 2005; Odom, Collet-Klingenberg, Rogers, & Hatton, 2010; Spooner et al., 2012). Again, those efforts emerged from the Division for Research of the CEC. A systematic process for identifying EBP in the field of special education is emerging based on past studies of EBP and recommendations from the Division for Research of the CEC. Cook and Cook (2013) identified four issues that are essential in the identification of EBP: "research design, quality of research, quantity of research, and magnitude of effect of supporting studies." According to Cook and Cook (2013), EBPs are identified through evidence-based reviews in which specific criteria are applied for each of these dimensions. Proposed standards for identifying EBP were established by the CEC Division for Research initiative discussed earlier; specifically, Gersten et al. (2005) identified standards for group and quasi-experimental research, and Horner et al. (2005) identified standards for single-subject research. Only recently (2014) has the umbrella organization of the Council for Exceptional Children published comprehensive guidelines for evaluating research to identify EBPs.

3.1.2 Research Design

Causality is typically inferred from high-quality group experimental, group quasi-experimental, and single-subject designs (Cook & Cook, 2013; Council for Exceptional Children, 2014; Gersten et al., 2005; Horner et al., 2005; Kratochwill et al., 2010; Odom et al., 2005). Regardless of the design, research must be of high quality to be trustworthy. Therefore, quality indicators for varying types of research must be used to determine if studies are of sufficient quality to be included as documentation for EBPs. For group designs, quality indicators based on Gersten et al. (2005), such as those developed by the National Professional Development Center on Autism Spectrum Disorders (NPDC-ASD; Wong et al., 2014), may be employed. For single-subject designs, quality indicators based on Horner et al. (2005) and/or Kratochwill et al. (2010) may be employed. Council for Exceptional Children (2014) recently published a comprehensive guide for identifying EBPs that includes quality indicators for group and single-subject designs.

3.1.3 Quantity of Research

EBPs can only be established when the findings of multiple high-quality research studies converge to provide sufficient evidence of the effectiveness of practices (Cook & Cook, 2013; Council for Exceptional Children, 2014). For the field of special education, Gersten et al. (2005) proposed that at least two high-quality or four acceptable-quality group experimental and/or quasi-experimental studies be required to establish EBPs. For single-subject designs, Horner et al. (2005) proposed that at least five high-quality studies from three different research groups in three different locations with a total of at least 20 participants be required to establish EBPs. When documenting the evidence for EBPs, some researchers have included a combination of evidence, such as requiring at least one high-quality experimental or quasi-experimental group design and at least three high-quality single-subject design articles conducted by at least two different research groups (Council for Exceptional Children, 2014; Wong et al., 2014).

3.1.4 Social Validity and Magnitude of Effect

Finally, EBPs must show positive effects on outcomes that are socially important (Cook & Cook, 2013; Council for Exceptional Children, 2014). Gersten et al. (2005) proposed that weighted effect sizes greater than zero from high-quality or adequate-quality studies are required for practices to be designated as EBPs. EBPs from high-quality single-subject research must be based on socially valid outcomes and quantitative indicators—five research studies by at least three different research groups with a minimum of 20 participants (Horner et al., 2005). EBP reviews should specify whether interventions produced positive, neutral, mixed, or negative effects, and CEC suggests that practices can be "classified in one of five categories: evidence-based practices, potentially evidence-based practices, mixed effects, insufficient evidence, or negative effects" (Council for Exceptional Children, 2014, p. 6).

3.1.5 EBPs for Students with Autism Spectrum Disorders

A good example of using quality indicators of research to identify EBP for students with autism spectrum disorders and of bridging the research to practice gap can be found in the NPDC-ASD (Odom et al., 2010; Wong et al., 2014). From the work of Odom and colleagues, approximately 27 EBPs have been identified for students with ASDs. Most of the practices are based on behavioral strategies that include prompting, reinforcement, extinction, antecedent-based intervention, discrete trial teaching, functional behavioral assessment and functional communication training, response interruption/redirection, self-management, and time delay. Other EBPs identified by the NPDC-ASD include exercise, modeling, naturalistic intervention, parent-implemented interventions, peer-mediated interventions, the Picture Exchange Communication System, pivotal response training, scripting, social narratives, social skills training, structured play groups, task analysis, technology-aided intervention, video modeling, and visual supports (Wong et al., 2014). Quality indicators based on Horner et al. (2005) and Gersten et al. (2005) were operationalized in a checklist-type format that was used to document that studies met quality indicators for research (Wong et al., 2014).

3.1.6 EBPs for Students with Severe Disabilities

Spooner et al. (2012) operationalized the Horner et al. (2005) criteria in a slightly different format to identify EBPs for teaching academics to students with severe developmental disabilities. Browder and Spooner and their colleagues have led the initiative to identify EBP for students with severe disabilities. By using quality indicators in systematic evidence-based reviews, they have identified constant time delay (Browder, Ahlgrim-Delzell, Spooner, Mims, & Baker, 2009; Spooner, Knight, Browder, Jimenez, & DiBiase, 2011; Spooner et al., 2012), task analytic instruction, and other systematic prompting (Spooner et al., 2011, 2012) as effective in teaching academic content to students with severe disabilities.

3.1.7 Implementing EBP

As noted by Odom et al. (2005), the identification of EBPs does not mean they will be used, or used with fidelity. Indeed, implementation research is required to identify processes by which the field can be motivated to adopt EBPs (Cook & Odom, 2013; Fixsen, Blasé, Horner, & Sugai, 2009). Nevertheless, the NPDC-ASD not only used quality indicators to identify EBPs for students with ASD but also used strategies from implementation science to try to effect change in actual practice (Odom et al., 2010). States had to complete an application process that required them to commit resources to a systems change initiative in which EBPs for students with ASD would be implemented with fidelity within model classrooms. Professional development was provided to teach practitioners to use EBPs, and teams within states used modeling, coaching, and technical assistance to support the implementation of EBPs with fidelity. Because parents of children with ASD are often effective advocates for their children, some states appeared to be motivated to implement EBP to address parents' concerns for high-quality interventions and to avoid litigation. Therefore, there may be more motivation for adopting EBP for students with ASD than there is for students with other types of disabilities.

3.2. Visual Impairment/Blindness

Consistent with Odom et al.'s (2005) and Cook and Odom's (2013) description of the challenges facing researchers in the field of special education, the field of visual impairments faces similar barriers. Specifically, the heterogeneity of students with visual impairments (ability levels, visual function, etc.), varying contexts of instruction (regular inclusive classes with support from itinerant TVIs, resource rooms, and classes in special schools), and complexity of identifying supports that are individualized for each student make the identification of EBPs for the field of visual impairment quite daunting. For example, students who are totally blind and functioning on grade level may receive beginning reading instruction within a regular class using the general education curriculum, with the TVI adapting all materials so that braille resources are similar to print resources that sighted students use. The TVI may then have individual time with the student for teaching the braille code. At a nearby school, a student with the same characteristics may be taught to read in a resource room using a special curriculum introducing braille contractions in a systematic manner. These differences in service delivery models would result from the individualized education program team's decisions.

In a 2006 review of intervention literature in the field of visual impairment, Ferrell reported that our field does not have any scientifically based practices based on criteria from the US Institute of Education Sciences' What Works Clearinghouse. The field has made limited progress in identifying prospective EBPs since that time. The articles commissioned for this special topics volume were solicited with the intention of identifying potential EBP for the field of visual impairment, with varying levels of success. Nevertheless, the quest must continue, and it can be informed by recent systematic efforts to identify effective practices both in the field of visual impairment and from other disability areas, as described earlier (Odom et al., 2010; Spooner et al., 2012; Wong et al., 2014).

As noted, Ferrell (2006) reviewed 40 years of peer-reviewed research in literacy and 50 years of research in mathematics in the field of visual impairment with the intention of using meta-analyses to identify effective practices meeting the highest standard of evidence from the What Works Clearinghouse. Most studies did not use designs that could demonstrate causality, and no studies had been replicated. Because of these issues, there were insufficient data for meta-analyses. Only 15 practices were identified as potential promising practices, and most of the studies from which the practices were derived were completed in the 1970s and 1980s; the quality of the research to support those 15 practices was not assessed.

Between 2007 and 2010, progress was made in reviewing, summarizing, critiquing, and evaluating the limited intervention literature for students with visual impairments published in peer-reviewed journals (Parker, Davidson, & Banda, 2007; Parker, Grimmett, & Summers, 2008; Parker & Pogrund, 2009; Wright, Harris, & Sticken, 2010). However, most of these reviews cannot be considered evidence-based reviews, as defined by

Cook and Cook (2013) and the Council for Exceptional Children (2014), because they did not include an examination of quality indicators for high-quality research and other criteria for establishing a practice as being evidence-based (e.g., restricting reviews to experimental or quasi-experimental group designs or single-subject designs, discussing magnitude of effect and social validity, and evaluating the quantity of research to determine if multiple research groups in multiple locations had replicated results with sufficient numbers of participants). Many of these authors initiated their reviews with the intention of completing meta-analyses using the criteria from the What Works Clearinghouse. However, they soon discovered a limited body of intervention research and found themselves summarizing the literature that was available rather than conducting meta-analyses to identify effective practices. These literature reviews have advanced our knowledge and understanding of the current state of intervention for students with visual impairments and document the need for high-quality research in the field. Indeed, as the chapters in this current publication attest, we have a very limited research foundation regarding practices that are effective for students with visual impairments.

3.2.1 Communication

In a review of intervention studies that did not include assessment of quality indicators of research, Parker et al. (2008) identified the Picture Communication Exchange System (PECS; Bondy & Frost, 1994; Frost & Bondy, 2002) as a potential effective strategy for students with visual impairments. Subsequently, Parker, Devender, Davidson, and Liu-Gitz (2009) adapted PECS for a student with visual impairment and autism in a single-subject design study. Later, Ivy, Hatton, and Hooper (2014) implemented a single-subject design study based on quality indicators and started accumulating evidence that adapted PECS may be used to help students initiate communication exchanges. PECS has been documented as an EBP for students with ASD (Odom et al., 2010; Wong et al., 2014), and adapted PECS (using tangible/tactile symbols instead of pictures) shows some evidence of being effective for students with visual impairments (Ivy et al., 2014; Parker et al., 2009); therefore, it seems appropriate to describe adapted PECS as a potential EBP for students with visual impairments.

Parker and Ivy (2014, this volume) did not evaluate quality indicators of the articles they reviewed for early communication interventions for children and youth with visual impairments; however, they did provide relatively detailed information about the studies they reviewed; this information could facilitate a more systematic investigation of some of the practices, such as communication partner training.

Earlier, Parker et al. (2007) reviewed 54 single-subject intervention research studies with participants who were deafblind, published between 1969 and 2006, but did not assess the quality of the research using quality indicators. Subsequently, Parker et al. (2008) examined 30 single-subject intervention studies, published between 1965 and 2006 that focused on improving communication skills in children with visual impairments and co-occurring disabilities, aged 6–22 years. Again, the authors did not assess the quality of the single-subject studies they reviewed; nevertheless, the identification of single-subject intervention studies and the categorization of the studies advanced our knowledge of potential interventions for facilitating early communication development. Specifically, Parker et al. (2008) reported that communication interventions for students with visual impairments/co-occurring disabilities fell into these categories: augmentative interventions involving microswitches, multicomponent partner training, dual communication boards, object symbol interventions, and adult-directed prompting and reinforcement.

Parker and Ivy (2014, this volume) expanded on the Parker et al. (2008) review by identifying 34 intervention research studies (21 with single-subject designs), published between 2000 and 2013, designed to promote early communication development in children with visual impairments, including deafblindness. These intervention studies were categorized as microswitch and technological supports ($n = 10$), multicomponent communication partner training programs ($n = 12$, with 9 conducted by the same research group), tangible symbols/adapted PECS ($n = 7$), narrative-/literacy-based interventions ($n = 3$), wait time ($n = 1$), and adapted prelinguistic milieu teaching ($n = 1$). Parker and Ivy (2014) extended previous reviews of communication research studies by providing a table with brief descriptions of outcomes, including the magnitude of effect. Even though Parker and Ivy (2014) did not assess quality indicators, their identification of intervention research for promoting communication in children with visual impairments, including those with deafblindness and other co-occurring disabilities, furthers our understanding of potentially effective interventions for promoting early communication. Now, if researchers can assess the quality, quantity, and social validity of this research, we may be able to identify potential EBPs for facilitating early communication development.

3.2.2 Assistive Technology

Kelly and Smith (2011) examined 45 years of literature on assistive technology with students with visual impairments published between 1965 and 2009. Their review focused on the use of assistive technology for classroom-based educational interventions with students with visual impairments, aged 3–21 years. Kelly and Smith (2011) identified two single-subject intervention studies as having adequate research designs to provide evidence of effectiveness: the use of a video magnifier on reading rates (LaGrow, 1981) and a comparison of the manual Perkins braillewriter and an electric braillewriter (Koenig & Ashcroft, 1983). Koenig and Ashcroft (1983) reported no differences in outcomes for the two methods of writing braille. Kelly and Smith (2011) did not evaluate these two studies using quality indicators, however.

More recently, Smith and Kelly (2014, this volume) reviewed literature on assistive technology interventions to improve educational outcomes from 2009 to 2013 and concluded that only two studies "included a research design with sufficient data to determine the effectiveness of an intervention with an intervention, control group, and comparison group." Although Kapperman, Sticken, and Smith (2011, 2012) described their studies as experimental, the studies appear to be quasi-experimental studies that used matched groups and pre- and posttests to assess the effectiveness of a software tutorial for a braille notetaker; the goal of the intervention was to improve knowledge of the Nemeth code for mathematics.

Although Smith and Kelly's (2014, this volume) review provides a helpful overview of the intervention literature on assistive technology in peer-reviewed publications, they included studies other than the experimental and quasi-experimental group designs and single-subject designs that have been deemed necessary for establishing causality and identifying EBP (Cook & Cook, 2013; Council for Exceptional Children, 2014). They also did not use quality indicators to assess the research, another requirement of identifying EBPs. Nevertheless, their identification of intervention research studies on the topic of AT is a valuable contribution to the field that highlights the need for rigorous intervention research in this area.

Smith and Kelly (2014, this volume) also documented the heterogeneity of AT intervention research that suggests that priorities for research on assistive technology for students with visual impairments have not been identified. Researchers interested in advancing EBP for the use of assistive technology with students with visual impairments may need to focus on designing intervention research that can establish causality to build the evidence base for effective practices, particularly in areas such as writing. Bickford and Falco's (2012) and Kamei-Hannan and Lawson's (2012) recent single-subject intervention research studies demonstrate the types of studies that are needed to advance academic outcomes such as writing and reading with assistive technology. However, research must meet standards for quality, quantity, social validity, and magnitude of effect to better inform our practice.

3.2.3 Systematic Prompting

Ivy and Hatton (2014, this volume) used in-depth assessment of quality indicators within and across single-subject research to evaluate systematic prompting. They found evidence to support systematic prompting to teach a variety of skills to a range of students who learn by nonvisual means; however, the field needs evidence for specific systematic prompting strategies (e.g., time delay) for specific skills (e. g., reading braille words) for specific types of students (e.g., tactile learners with intellectual disabilities) within a variety of contexts (e.g., specialized school for students with visual impairments). It does appear, however, that systematic prompting to teach a variety of skills (as discussed in Ivy & Hatton, 2014, this volume) to tactile learners is the only EBP, to date, that has been identified for students with visual impairments based on assessment of quality and quantity indicators of research. Currently, we have the most evidence for the system of least prompts, followed by most to least prompting systems, and then graduated guidance to teach chained skills to students whose multiple disabilities include visual impairment. We now need to strengthen the evidence using well-planned and well-executed research by multiple researchers in our field.

3.2.4 Literacy

Along with Ferrell (2006), Erickson and Hatton (2007) identified repeated reading as a potentially effective practice for increasing reading fluency and comprehension of students with low vision. Specifically, Layton and Koenig's (1998) intervention using a single-subject design across four students with low vision may be a promising practice for literacy instruction with children with visual impairments. Repeated reading has been found to be effective in improving fluency and comprehension in students with learning disabilities and with nondisabled students (Therrien, 2004).

Savaiano and Hatton (2013) replicated Layton and Koenig's (1998) study and incorporated quality indicators of single-subject research. A variation of Layton and Koenig's (1998) procedure was used by Pattillo, Heller, and

Smith (2004) with three students with low vision and with one braille reader who used optical character recognition software for repeated readings. Consequently, three different research groups have used repeated reading with a total of 12 students (one braille reader) to improve reading speed and/or comprehension. However, Layton and Koenig (1998) and Pattillo et al. (2004) were published before quality indicators were available for rating single-subject studies. Even so, we seem to have emerging evidence that repeated reading may be a potential EBP for students with low vision who read print. Repeated reading is relatively easy to implement and can enhance fluency and comprehension; therefore, it should be relatively easy for the field to establish it as an EBP.

Although Parker and Pogrund (2009) reviewed research literature on literacy for students with visual impairments and co-occurring disabilities, seven of the nine studies they identified used quality case study designs, one used a correlational design, and the one study that used single-subject design focused on the use of embedded instruction in general education classrooms (Johnson, McDonnell, Holzwarth, & Hunter, 2004) and was published before quality indicators were available for single-subject research.

Savaiano, Compton, and Hatton's (2014, this volume) recent review of the literature on reading comprehension interventions for braille readers was consistent with earlier reviews of literacy for students with visual impairments (Erickson & Hatton, 2007; Ferrell, 2006; Parker & Pogrund, 2009); we have very limited research to guide practice. Specifically, of the 12 studies in 11 articles that were reviewed, 11 were correlational descriptive studies, and the results from the one single-subject design study showed that the student reached criterion before the intervention was implemented. Although Savaiano et al. used quality indicators to rate the studies that were included, correlational research cannot be used to establish causality. However, the findings from Savaiano et al. (2014) advance our understanding of factors that may impact reading comprehension in readers of braille and provided a potential framework for completing systematic intervention research to identify EBPs for braille readers.

3.2.5 Optical Devices to Increase Reading Speed and Fluency

Although the use of optical devices to increase reading speed, comprehension, and efficiency was identified as a promising practice for students with low vision (Ferrell, 2006), the studies upon which this assessment was made have primarily been correlational descriptive studies (Corn, Wall, & Bell, 2000; Corn et al., 2002; Lackey, Efron, & Rowls, 1982). Researchers comparing reading speed, comprehension, and/or efficiency of students using optical devices with regular print to those who use large print text have not reported that one medium was more effective than another (Corn et al., 2002; Lussenhop & Corn, 2002; Morris, 1973; Sykes, 1971).

Savaiano (2011) identified four studies on the use of optical devices to increase reading speed that used either pre-/posttest group (Corn, Wall, et al., 2000; Corn et al., 2002; Farmer & Morse, 2007) or single-subject design (Smith & Erin, 2002) and then used quality indicators (Gersten et al., 2005; Horner et al., 2005) to evaluate the studies. Most of the studies were published before quality indicators of research were developed; therefore, it was not surprising that most of them failed to meet those standards. Only one of the group design studies appeared to have sufficient statistical power to detect effect sizes of less than 0.8 (Corn et al., 2002).

It appears that individual preference may be the key factor in determining the most effective accommodation for readers with low vision based on recent single-subject research (Hanner, 2012; Lusk, 2012). The importance of optical devices to provide access to print for students with low vision is unquestionable; however, individual differences and preferences may make research to establish use of optical devices as an EBP a moot point. All students with low vision whose clinical low vision evaluation documents the need for optical devices should have the devices. The issue of large print versus regular print/optical devices has been inconclusive, probably due to individual differences. It may be most helpful to identify the most efficient strategies to teach students with low vision to use their optical devices and to demonstrate the importance of introducing optical devices early.

3.2.6 Mathematics

Limited information is available regarding potentially effective practices for teaching mathematics. Although Ferrell (2006) reported effect sizes for some math interventions, quality indicators were not used to evaluate the research, and none of the studies had been replicated. Nevertheless, there was some indication that concrete manipulative and assistive technology may enhance math instruction (Ferrell, 2006). Although Smith and Kelly (2014, this volume) identified studies of a math software tutorial for the Nemeth code as the only resource with sufficient evidence to be considered an effective practice, they did not use quality indicators to evaluate the study. In fact, very little information was provided about the intervention, making it impossible for others to replicate the study. The authors state that the intervention was a software tutorial on the Nemeth math code; however, no information was provided about the actual content, the frequency and length of training sessions, and the fidelity of implementation. No proof was provided that students actually completed the tutorial. The authors state that an experimental design was used; however, it appears that a quasi-experimental pre-/posttest group design was used. Students were matched on a number of characteristics; however, no descriptive data or statistical data were provided to document that the groups were equally matched.

4. IMPLICATIONS FOR THE FUTURE

In the past decade, a number of reviews to identify effective practices for educating students with visual impairments have been published. Although many of the reviews were broad and did not assess the quality of research being reviewed, they have advanced our understanding of the limited intervention research available to guide practice for students with visual impairments. These past reviews also documented the fact that most studies cannot be used to identify EBP in our field because they did not employ experimental or quasi-experimental group designs or high-quality single-subject designs. In situations in which appropriate designs were used, research was typically not evaluated using the systematic process required to identify EBPs (Cook & Cook, 2013; Council for Exceptional Children, 2014; Gersten et al., 2005; Horner et al., 2005; Kratochwill et al., 2010). Even so, these earlier reviews identified research that had been conducted and advanced our knowledge of potential educational interventions.

Ivy and Hatton's (2014, this volume) in-depth assessment of quality indicators of single-subject research on systematic prompting may represent the most rigorous assessment of the research literature on a specific practice for students who are tactile learners that has been published to date. However, there are four categories of systematic prompting that have been used to teach a variety of skills to students with varying levels of ability. We need to establish evidence for each of the four categories—time delay, most to least prompts, least to most prompts, and graduated guidance—for students with low vision and blindness. Such research might reveal that most to least prompts are more appropriate for teaching self-help skills to students who are primarily tactile learners, while least to most prompts or graduated guidance would be more efficient for students who are visual learners with relatively good low vision. Time delay was used to teach braille readers with intellectual disabilities highly motivating words (Hooper, Ivy, & Hatton, 2014), and so, it would probably be effective to teach sight words to students with low vision and intellectual disabilities. Browder et al. (2009) had already established the evidence base for time delay to teach literacy skills to students with severe disabilities, and so, it seems likely that it would be effective with students with low vision and intellectual disabilities. At a minimum, the process that Ivy and Hatton (2014, this volume) used to evaluate single-subject research can serve as a model for future reviews of single-subject studies.

4.1. Advancing EBP Through Research

More importantly, though, the field needs to design and implement high-quality research to identify practices that can improve student outcomes. Because of the low-incidence nature of visual impairment, single-subject studies are probably the most practical design. With instruction and support, teachers and students enrolled in personnel preparation programs can conduct single-subject research, particularly if they partner with colleagues in university programs with expertise in this area. Professional development may be needed to assure that the studies that we design are of the highest possible quality so that they meet quality indicators for single-subject research (Horner et al., 2005; Kratochwill et al., 2010). Collaborative research groups such as those proposed by Smith and Kelly (2014, this volume) may be better able to design and implement high-quality experimental or quasi-experimental group research designs.

4.2. Advancing EBP Through Personnel Preparation

Personnel preparation programs can advance the identification and use of EBP by integrating readings on this topic and by requiring students to complete theses and other high-stakes assignments that advance the identification of EBP. Since 2011, a major goal of the Peabody College's Program in Visual Disabilities at Vanderbilt University has been to advance the use of EBP with students with visual impairments. Peabody College of Vanderbilt University has prepared TVIs since the 1950s when the Department of Special Education was created. As early as the 1920s, Peabody College offered support to teachers and families of students with visual impairments through summer courses and workshops. A major characteristic of the Peabody College's Department of Special Education, within which the program is housed, is intervention research using single-subject methods.

We conducted a review of the literature in summer 2011 to identify resources on EBP for our courses. Subsequently, our doctoral and masteral students conducted research on the use of repeated reading, the *Picture Exchange Communication System* (PECS; Frost & Bondy, 2002) using tangible symbols, constant time delay, self-monitoring, and the use of optical devices. Three of these studies resulted in publications, with the anticipated publication of two other studies that are being prepared (Hooper et al., 2014; Ivy et al., 2014; Savaiano & Hatton, 2013). Departmental grants of less than $1000 per study were the only financial support available to support this research.

In addition to preparing TVIs who can improve their students' outcomes (career and college readiness and quality of life), personnel preparation programs should prepare practitioner scientists who will use EBPs and then share their experiences with the field through publications and presentations. Students become much more aware of professional organizations when they prepare proposals for presentations and attend state and national conferences.

Using research literature that scholars locate each semester in each course, the field can continuously identify promising practices to inform intervention and future research. Scholars can operationalize these practices using task analysis before using them in field experiences. Along with task analyses, future teachers can create fidelity of implementation checklists so they can assess each others' capacity to implement practices with accuracy. Spooner et al. (2012) identified the use of task analysis as an EBP in teaching academics to students with severe disabilities, and Odom and colleagues (Odom et al., 2010; Wong et al., 2014) identified task analysis as an EBP for students with ASD. Scholars can become more efficient with the process of task analysis if they are required to develop task analyses based on high-quality research studies. They can translate research articles about EBP to teacher-friendly, task-analyzed steps for implementing the practice; implement the promising practices with fidelity; and use data-driven decision making in their fieldwork.

4.3. Advancing EBP Through Collaboration

Many of the challenges impeding research in our field can be addressed through collaboration among personnel preparation programs, specialized schools for students with visual impairments, local school systems, consumer/advocacy groups, and families. Through collaboration, we can identify the most critical research priorities so that research is socially valid. Schools may welcome researchers who target critical issues such as improved reading and writing scores for high-stakes assessments. Families will be more motivated to participate in research if they have identified research priorities that are important to their children and themselves. By having frequent and open communication, these collaborators can address constraints and resources that will facilitate research to improve outcomes for students with visual impairments and their families. At the same time, future teachers will acquire the skills they need to be practitioner scientists and to collect high-quality progress-monitoring data. Because their evaluations as university students and future teachers will probably be based, in part, on their own students' outcomes on high-stakes assessments, there has never been a better time to advance the use of effective practices in the education of students with visual impairments.

REFERENCES

Ajuwon, P. M., & Oyinlade, A. O. (2008). Educational placement of children who are blind or have low vision in residential schools and public schools: A national study of parents' perspectives. *Journal of Visual Impairment & Blindness, 102*, 325–339.

Bickford, J. O., & Falco, J. R. (2012). Technology for early braille literacy: Comparison of traditional braille instruction and instruction with an electronic notetaker. *Journal of Visual Impairment & Blindness, 107*, 679–693.

Bondy, A. S., & Frost, L. A. (1994). The picture exchange communication system. *Focus on Autistic Behavior, 9*, 1–19.

Browder, D., Ahlgrim-Delzell, L., Spooner, F., Mims, P. J., & Baker, J. N. (2009). Using time delay to teach literacy to students with severe developmental disabilities. *Exceptional Children, 75*(3), 343–364.

CAEP Commission on Standards and Performance Reporting. (2013). *Accreditation standards and recommendations*. Retrieved from, http://caepnet.files.wordpress.com/2013/02/draft_standards3.pdf.

Chudowsky, N., Chudowsky, V., & Keber, N. (2009). *State test score trends through 2007–08, part 4: Has progress been made in raising achievement for students with disabilities?* Washington, DC: Center on Education Policy. Retrieved January 15, 2012 from, http://www.edweek.org/media/studentswdisabilities.pdf.

Cook, B. G., & Cook, S. C. (2013). Unraveling evidence-based practices in special education. *Journal of Special Education, 47*(2), 71–82. http://dx.doi.org/10.1177/0022466911420877.

Cook, B. G., & Odom, S. L. (2013). Evidence-based practices and implementation science in special education. *Exceptional Children, 79*(2), 135–144. Retrieved from, http://search.proquest.com.proxy.library.vanderbilt.edu/docview/1270781695?accountid=14816.

Corn, A. L., DePriest, L. B., & Erin, J. N. (2000). Visual efficiency. In A. Koenig & M. Holbrook (Eds.), *Foundations of education: Instructional strategies for teaching children and youths with visual impairments*: *Vol. 2*. (pp. 464–499) (2nd ed.). New York, NY: AFB Press.

Corn, A. L., & Koenig, A. J. (1996). Introduction. In A. L. Corn, & A. J. Koenig (Eds.), *Foundation of low vision: Clinical and functional perspectives*. New York, NY: AFB Press.

Corn, A. L., & Lusk, K. (2010). Perspectives on low vision. In A. Corn, & J. Erin (Eds.), *Foundations of low vision: Clinical and functional perspectives* (pp. 3–34) (2nd ed.). New York, NY: AFB Press.

Corn, A., Wall, R., & Bell, J. (2000). Impact of optical devices on reading rates and expectations for visual functioning of school-age children and youth with low vision. *Visual Impairment Research, 2*(1), 33–41.

Corn, A. L., Wall, R. S., Jose, R. T., Bell, J. K., Wilcox, K., & Perez, A. (2002). An initial study of reading and comprehension rates for students who received optical devices. *Journal of Visual Impairment & Blindness, 96*(5), 322–334.

Council for Exceptional Children (2014). *Standards for evidence-based practices in special education.* Retrieved from, www.cec.sped.org/media/Files/Standards/EvidencebasedPracticesandPractice/CECsEBPStandards.pdf.

Council of Chief State School Officers. (2012). *Our responsibility, our promise: Transforming educator preparation and entry into the profession.* Retrieved from, http://www.ccsso.org/Documents/2012/Our%20Responsibility%20Our%20Promise_2012.pdf.

Education for All Handicapped Children Act of 1975, 20 U.S.C. § 1400 et seq. (1975).

Erickson, K. A., & Hatton, D. D. (2007). Literacy and visual impairment. *Seminars in Speech and Language, 28*, 58–68.

Elementary and Secondary Education Act, Pub. L. No. 107–110, § 115, Stat. 1425 (2002).

Farmer, J., & Morse, S. E. (2007). Project magnify: Increasing reading skills in students with low vision. *Journal of Visual Impairment & Blindness, 101*(12), 763–768.

Ferrell, K. A. (2006). Evidence-based practices for students with visual disabilities. *Communication Disorders Quarterly, 28*, 42–48.

Fixsen, D. L., Blasé, K. A., Horner, R., & Sugai, G. (2009). *Scaling up evidence-based practices in education. Scaling up Brief # 1.* Chapel Hill: The University of North Carolina, FPG, SISEP.

Frost, L., & Bondy, A. S. (2002). *The picture exchange communication system training manual* (2nd ed.). Newark, DE: Pyramid Educational Products.

Gersten, R., Fuchs, L., Compton, D., Coyne, M., Greenwood, C., & Innocenti, M. S. (2005). Quality indicators for group experimental and quasi-experimental research in special education. *Exceptional Children, 71*, 149–164.

Hanner, L. A. (2012). *Magnification or large print: A single subject comparison.* Unpublished master's thesis. Vanderbilt University, Nashville, TN.

Harr-Robins, J., Song, M., Hurlburt, S., Pruce, C., Danielson, L., Garet, M., et al. (2012). *The inclusion of students with disabilities in school accountability systems. Interim Report NCEE 2012-4056.* Washington, DC: National Center for Education Evaluation and Regional Assistance, Institute of Educational Science, U.S. Department of Education.

Hatlen, P. (2000). Historical perspectives. In M. C. Holbrook, & A. J. Koenig (Eds.), *Foundations of education: History and theory of teaching children and youths with visual impairments: Vol. 1.* (pp. 1–54) (2nd ed.). New York, NY: AFB Press.

Hooper, J. D., Ivy, S. E., & Hatton, D. D. (2014). Using time delay to teach automatic word recognition to braille students. *Journal of Visual Impairment & Blindness, 108*, 107–121.

Horner, R. H., Carr, E. G., Halle, J., McGee, G., Odom, S., & Wolery, M. (2005). The use of single-subject research to identify evidence-based practice in special education. *Exceptional Children, 71*, 165–179.

Huebner, K. (2000). Visual impairment. In M. C. Holbrook & A. J. Koenig (Eds.), *Foundations of education: History and theory of teaching children and youths with visual impairments: Vol. 1.* (pp. 55–76) (2nd ed.). New York, NY: AFB Press.

Individuals with Disabilities Education Improvement Act of 2004, PL 108–466, 20 U. S. C. §1400, H. R. 1350.

Ivy, S. E., & Hatton, D. D. (2014). Teaching skill acquisition to individuals with blindness: A systematic review of response prompting procedures. *International Review of Research in Developmental Disabilities, 46*, 55–100.

Ivy, S. E., Hatton, D. D., & Hooper, J. D. (2014). Using the picture exchange communication system with students with visual impairment. *Exceptional Children.* In press.

Johnson, J. W., McDonnell, J., Holzwarth, V. N., & Hunter, K. (2004). The efficacy of embedded instruction for students with developmental disabilities enrolled in general education classes. *Journal of Positive Behavior Interventions, 6*(4), 214–227.

Kamei-Hannan, C., & Lawson, H. (2012). Impact of a Braille-note on writing: Evaluating the process, quality, and attitudes of three students who are visually impaired. *Journal of Special Education Technology, 27*, 1–14.

Kapperman, G., Sticken, J., & Smith, T. J. (2011). The effectiveness of the Nemeth code tutorial for the BrailleNote. *Insight: Research and Practice in Visual Impairment and Blindness, 4*, 22–28.

Kapperman, G., Sticken, J., & Smith, T. J. (2012). The effectiveness of the Nemeth code tutorial for the BrailleNote: A follow-up study. *Insight: Research and Practice in Visual Impairment and Blindness, 5*, 168–174.

Kelly, S. M., & Smith, D. W. (2011). The impact of assistive technology on the educational performance of students with visual impairments: A synthesis of research. *Journal of Visual Impairment & Blindness, 105*(2), 73–83.

Koenig, A., & Ashcroft, S. C. (1983). Initial assessment of student use of the electric Perkins Brailler. *Education of the Visually Handicapped, 15*(2), 59–66.

Koenig, A. J., & Holbrook, M. C. (2000). Ensuring high-quality instruction for students in braille literacy programs. *Journal of Visual Impairment & Blindness, 94*, 677–694.

Koretz, D. (2008). A measured approach: Value-added models are a promising improvement, but no one measure can evaluate teacher performance. *American Educator, 39*, 18–27.

Kratochwill, T. R., Hitchcock, J., Horner, R. H., Levin, J. R., Odom, S. L., Rindskopf, D. M., et al. (2010). *Single-case designs technical documentation.* Retrieved from What Works Clearinghouse website, http://ies.ed.gov/ncee/wwc/pdf/wwc_scd.pdf.

Lackey, G. H., Efron, M., & Rowls, M. D. (1982). For more reading: Large print books or the Visolett? *Education of the Visually Handicapped, 14*(3), 87–94.

LaGrow, S. J. (1981). Effects of training on CCTV reading rates of visually impaired students. *Journal of Visual Impairment & Blindness, 75*, 368–375.

Layton, C. A., & Koenig, A. J. (1998). Increased reading fluency in elementary students with low vision through repeated readings. *Journal of Visual Impairments & Blindness, 92*, 276–292.

Lusk, K. E. (2012). The effects of various mounting systems of near magnification on reading performance and preference in school-age students with low vision. *British Journal of Visual Impairment, 30*(3), 168–181.

Lussenhop, K., & Corn, A. L. (2002). Comparative studies of the reading performance of students with low vision. *RE:view, 34*(2), 57–69.

Misco, T. (2008). Was that a result of my teaching? A brief exploration of value-added assessment. *Clearing House, 82*(1), 11–14.

Morris, O. (1973). *Reading performance of normally sighted and partially sighted third and fourth grade students using regular print and large print.* Unpublished doctoral dissertation, University of Minnesota, Minneapolis.

No Child Left Behind Act of 2001, 20 U.S.C. 70 § 6301 et seq. (2002).

Odom, S. L., Brantlinger, E., Gersten, R., Horner, R. H., Thompson, B., & Harris, K. (2005). Research in special education: Scientific methods and evidence-based practices. *Exceptional Children, 71*(2), 137–148.

Odom, S. L., Collet-Klingenberg, L., Rogers, S. J., & Hatton, D. D. (2010). Evidence-based practices in interventions for children and youth with autism spectrum disorders. *Preventing School Failure, 54*(40), 275–282.

Parker, A. T., Davidson, R., & Banda, D. R. (2007). Emerging evidence from single-subject research in the field of deaf-blindness. *Journal of Visual Impairment & Blindness, 101*(11), 690–700.

Parker, A. T., Devender, B. R., Davidson, R. C., & Liu-Gitz, L. (2009). Adapting the picture exchange communication system for a student with visual impairment and autism: A case study. *Research and Practice in Visual Impairment & Blindness*, *3*(1), 2–11.

Parker, A. T., Grimmett, E. S., & Summers, S. (2008). Evidence-based communication practices for children with visual impairments and additional disabilities: An examination of single-subject design studies. *Journal of Visual Impairment & Blindness*, *102*(9), 540–552.

Parker, A., & Ivy, S. (2014). Communication development of children with visual impairments: A synthesis of intervention research. *International Review of Research in Developmental Disabilities*, *46*, 101–144.

Parker, A. T., & Pogrund, R. L. (2009). A review of research on the literacy of students with visual impairments and additional disabilities. *Journal of Visual Impairment & Blindness*, *103*(10), 635–648.

Pattillo, S. T., Heller, K. W., & Smith, M. (2004). The impact of a modified repeated-reading strategy paired with optical character recognition on the reading rates of students with visual impairments. *Journal of Visual Impairment & Blindness*, *98*, 1.

Savaiano, M. E. (2011). *Effects of training and use of optical devices on reading speed of children with low vision.* Unpublished manuscript, Department of Special Education, Vanderbilt University, Nashville, TN.

Savaiano, M. E., Compton, D. L., & Hatton, D. D. (2014). Reading comprehension for braille readers: An empirical framework for research. *International Review of Research in Developmental Disabilities*, *46*, 177–206.

Savaiano, M. E., & Hatton, D. D. (2013). Improving reading speed and comprehension of students with visual impairments. *Journal of Visual Impairment & Blindness*, *107*, 93–106.

Smith, J. K., & Erin, J. N. (2002). The effects of practice with prescribed reading glasses on students with low vision. *Journal of Visual Impairment & Blindness*, *96*, 765–782.

Smith, D. W., & Kelly, S. M. (2014). Assistive technology for students with visual impairments: A research agenda. *International Review of Research in Developmental Disabilities*, *46*, 23–54.

Spooner, F., Knight, V., Browder, D., Jimenez, B., & DiBiase, W. (2011). Evaluating evidence-based practice in teaching science content to students with severe developmental disabilities. *Research and Practice for Persons with Severe Disabilities*, *36*(1–2), 1–2.

Spooner, F., Knight, V. F., Browder, D. M., & Smith, B. R. (2012). Evidence-based practice for teaching academics to students with severe developmental disabilities. *Remedial and Special Education*, *33*(6), 374–387. http://dx.doi.org/10.1177/0741932511421634.

Sykes, K. S. (1971). A comparison of the effectiveness of standard print in facilitating the reading of visually impaired students. *Education of the Visually Handicapped*, *3*, 97–106.

Therrien, W. J. (2004). Fluency and comprehension gains as a result of repeated reading: A meta-analysis. *Remedial and Special Education*, *25*, 252. http://dx.doi.org/10.1177/07419325040250040801.

Wong, C., Odom, S. L., Hume, K., Cox, A. W., Fettig, A., Kucharczyk, S., et al. (2014). *Evidence-based practices for children, youth, and young adults with autism spectrum disorder.* Chapel Hill: The University of North Carolina, Frank Porter Graham Child Development Institute, Autism Evidence-Based Practice Review Group.

Wright, T., Harris, B., & Sticken, E. (2010). A best-evidence synthesis of research on orientation and mobility involving tactile maps and models. *Journal of Visual Impairment & Blindness*, *105*, 95–106.

CHAPTER TWO

Assistive Technology for Students with Visual Impairments: A Research Agenda

Derrick W. Smith[*,1], **Stacy M. Kelly**[†]
[*]Department of Education, University of Alabama in Huntsville, Huntsville, Alabama, USA
[†]Department of Special and Early Education, Northern Illinois University, DeKalb, Illinois, USA
[1]Corresponding author: e-mail address: derrick.smith@uah.edu

Contents

Abstract

This research synthesis examined the literature on assistive technology in education for individuals with visual impairments from September 2009 to May 2013 and is a replication of a previous study by Kelly and Smith (2011). To identify changes in the types, methodology, and rigor of research since August 2009, we located and analyzed 141 articles in English peer-reviewed publications. Only 2 of the 141 articles met the standards for scientific research developed by the What Works Clearinghouse. While only two studies met criteria for group research, there was an increase in the quantity of published articles, including a number of qualitative and single-subject research

International Review of Research in Developmental Disabilities, Volume 46
ISSN 2211-6095
http://dx.doi.org/10.1016/B978-0-12-420039-5.00003-4

studies, suggesting that more and higher quality research is being conducted. A succinct summary of notable studies is provided as well as possible strategies to advance research on assistive technology for students with visual impairments to promote more rigorous and robust standards/competencies.

1. INTRODUCTION

Technology has come a long way in a remarkably short period of time and has become a driving force in society. Mobile devices, such as smart phones, tablet computers, and personal digital assistants are used everywhere in the modern world. These advancements have major implications for people who are visually impaired (i.e., those who are blind or have low vision). The theoretical framework presented by Lowenfeld (1973) captures the essence of this impact.

Lowenfeld (1973) determined that there are three primary issues facing individuals with visual impairments: access to information, independent travel, and a lack of meaningful experiences. Assistive technology (AT) can help to resolve each of these limitations in substantial ways. No longer do people who are blind have to wait for someone else to transcribe ordinary printed text into braille for them. People who are blind can read online newspapers, check email messages, connect to the Internet, navigate favorite tablet computer applications, or download documents, thus having instant access to information available to sighted peers. Individuals who are visually impaired also use AT, such as specifically designed global positioning systems, talking maps, and screen reading software for accessing computers for information to travel independently. The combination of access to information and independent travel provides opportunities for meaningful experiences and for digital social opportunities (e.g., Facebook or LinkedIn) for forming meaningful friendships and participating in online collaboration (Kelly & Smith, 2008).

Technology has revolutionized classrooms and learning tools used by students with visual impairments. Students who are blind can download, print, and submit their homework to their classroom teachers at the same time as their sighted peers without requiring assistance from others. Students with low vision can use digital cameras to read print classroom materials and simply move the digital camera to access information from the board across the room. AT removes barriers to student performance and enables students who are visually impaired to maintain or improve their quality of life.

1.1. Complexity of Using AT with Students with Visual Impairment

While technology is considered the "great equalizer" for individuals with disabilities (Michaels & McDermott, 2003), the potential impact of technology may not be realized (Johnson & Evans, 2005). Research on the educational impact of the use of technology (such as computers and the Internet) has reported mixed results. Findings from some studies have supported the use of these technologies in all or certain academic areas, such as reading or math (Attewell & Battle, 1999; Jackson et al., 2006; Judge, Puckett, & Bell, 2006), while other findings were less favorable (Hunley et al., 2005; Wittwer & Senkbeil, 2008). As will be determined, the findings for the use of technology for students with visual impairments are no less complicated.

There is extraordinary complexity involved in educating students with visual impairments through AT. The field of visual impairments is unique in that the population is incredibly diverse. At least half of all students with visual impairments have additional disabilities (Hatton, Ivy, & Boyer, 2013; Sacks, 1998). For example, individuals may be academically or intellectually gifted and visually impaired, or individuals may be severely disabled and visually impaired. The range of additional disabilities experienced by any one individual who is visually impaired influences the way in which information is acquired and learned for that person (Sacks, 1998). Consider the impact of hearing impairments, physical disabilities, learning disabilities, traumatic brain injury, or autism on the learning of students who are also visually impaired. Due to the complexity of the students' disabilities, they will most likely require a diverse set of educational, programmatic, and technological components in their programs.

There is also wide variation in the severity of visual impairment. For example, individuals with a specific visual condition such as retinopathy of prematurity may be totally blind and see nothing at all or have minimal residual vision that hinders completion of daily activities. Another individual with the same diagnosis may have enough residual vision to use vision as their primary learning medium. Furthermore, it is likely that no two people with visual impairments use their remaining vision in the exact same way (Corn & Lusk, 2010). Because students with visual impairments are so heterogeneous, they require a range of educational placements that adds to the complexity of the appropriate use of AT to meet a wide range of educational and daily living goals.

1.2. Previous Reviews on Assistive Technology for Students with Visual Impairment

In 2011, Kelly and Smith synthesized the literature on AT intended for students with visual impairments published from 1965 to 2009. They located 256 articles on the topic but identified only two promising evidence-based practices (Koenig & Ashcroft, 1983; LaGrow, 1981). As noted in the article, the lack of scientifically based research on the impact of AT on learning was a major issue for teachers of students with visual impairments and their students. The lack of rigorous research limits teachers' ability to identify specific tools for the specific needs of their students. Further confounding these challenges is the No Child Left Behind Act of 2001 (NCLB, 2002, 20 USC § 1401, § 300.5) and the Individuals with Disabilities Education Improvement Act (IDEIA, P.L. 108-446) (Individuals with Disabilities Education Improvement Act, 2004) that require the use of scientifically based practices, interventions, and treatments for students with disabilities (Peterson-Karlan & Parette, 2007). Regardless of the potential impact of AT on students, educators must be able to base their AT practices on research that currently may not exist.

Kelly and Smith (2011) challenged the fields of AT and visual impairments to "begin to replicate studies and to ensure that the studies follow even higher standards" (p. 81). They urged the field to continually complete rigorous research on emerging technologies to provide evidence for educators to justify the use of such resources. To extend earlier research, the goal of this study is to replicate our previous work to determine if there were any notable differences in the literature on AT for students with visual impairments since 2009.

Replication of research is a literal repetition of previous research in order to synthesize and enhance the inferences drawn from the research combined (Schafer, 2001). This study set out to replicate Kelly and Smith's (2011) earlier synthesis that examined the literature on educational AT for students with visual impairments, aged 3–21 years, from January 1965 to August 2009. Specifically, we examined the same topic using articles published between September 2009 and May 2013. The same research questions (Kelly & Smith, 2011, p. 75) were used for this investigation: "What is our knowledge basis for assistive technology used in the education of school-aged individuals with visual impairments? To what extent has the effectiveness of assistive technology used by school-aged individuals with visual impairments been determined through scientifically-based research methods?"

2. METHODS

2.1. Definitions

This study investigated these two broad research questions using the federal definition of AT. According to the IDEIA (P.L. 108-446), AT (devices) are "any item, piece of equipment, or product system, whether acquired commercially off the shelf, modified, or customized, that is used to increase, maintain, or improve functional capabilities of individuals with disabilities" (1997, Sec. 602, 20 USC 1401, § 300.5).

For more than a decade in the United States, federal initiatives have underscored the importance of "evidence-based" decision making for education and early intervention. For example, the What Works Clearinghouse (WWC) was established in 2002 by the US Department of Education as a source for evidence-based practices in education to improve outcomes for students. The Council for Exceptional Children Division for Research (CEC-DR, 2004) led efforts within special education to determine the types of research that could be considered high quality and lead to evidenced-based practices.

An assessment of the effectiveness of evidence-based practices for students with low vision and blindness can be quite challenging within this conceptual framework of credible and reliable evidence (Kelly & Smith, 2011). The devices and software of interest cannot be deemed effective simply because there are well-known practical applications (Kelly & Smith, 2011). Therefore, we used the following definitions to guide our examination and critique of research on AT for students with visual impairments. *Effective* was defined as "having a positive impact on education" and *effectiveness* was defined as "the degree to which assistive technology had a positive impact on educational performance" (Kelly & Smith, 2011, p. 74). The definition of *evidence-based research* was established as "the use of scientific research to establish best practices based on evaluation of research," while *scientifically based research* was defined as "a research design that determines with the highest degree of probability whether intervention caused the effects" (Kelly & Smith, 2011, p. 74).

2.2. Inclusion and Exclusion Criteria

We applied the federal definition of AT to classroom-based interventions. That is, we only considered AT research that had educational and classroom-based applications. Specifically, we examined literature on AT devices and software that provide students with visual impairments access to the general curriculum and direct instruction in areas of the expanded core curriculum within school classrooms. We examined literature published between September 2009 and May 2013. Consistent with Kelly and Smith (2011), we included studies that (a) focused on the use of AT for classroom-based education and (b) included participants with blindness or low vision, regardless of co-occurring disabilities, between the ages of 3 and 21 years. Also consistent with Kelly and Smith (2011), the methodology excluded studies that focused on the use of AT "for orientation and mobility, independent living, or augmentative and alternative communication" (p. 39).

2.3. Search Procedures

Five electronic databases were searched using the Boolean search method: EBSCO Education Full-Text, ERIC, Proquest Education Journals, Psychinfo, and Psych Articles. To define the search, the Boolean operator "and" was used, while the "not" operator was used to exclude specific terms outside the purview of this study. Articles were retrieved if they included the search terms specified and none of the excluded terms.

The specific search terms used in each of the five databases were *blind, deaf blind, deafblind, deaf-blind, eye disorders, partially sighted, vision disorders, visual disabilities, blindness, visual impairment, visual impairments, visually impaired, partial vision,* and *large print,* each paired with the terms *assistive technology, instructional technology, assistive device, communication devices, accessibility, technology, computers, electronic aids, optical aids, notetaker, low vision aids, tablet, cell phone, mobile, iOS, app, smartphone, iPad, iPod, iPhone, Android, laptop, haptic touch, interactive whiteboards,* and *refreshable braille display.* The search terms excluded articles with the following terms in the abstract or title: *AAC, augmentative device, augmentative communication, alternative communication, augmentative and alternative communication, orientation and mobility, O&M, or COMS.*

Following the initial search of electronic databases, we conducted an ancestral search using the same parameters with the articles retrieved from the database search. We also used a manual search for articles on AT in *The Journal of Visual Impairments & Blindness (JVIB)*, *Insight: Research and Practice in Visual Impairment and Blindness* (named *AER Journal: Research and Practice in Visual Impairment and Blindness* prior to 2011), *the Journal of Blindness Innovation and Research (JBIR)*, and *the British Journal of Visual Impairment.*

In the final step in the search process, we examined the reference lists of all articles and additional articles that met the general search criteria were retrieved (Kelly & Smith, 2011).

The articles identified for inclusion in the study were classified into the same nine distinct categories Kelly and Smith (2011) used to present the results of the synthesis of the AT literature from earlier time periods. Then the articles that used a research methodology (experimental, correlational, descriptive, single-subject, and qualitative) were categorized further using the Council for Exceptional Children Division for Research (2004) conceptual groupings for evidence-based research. For further clarification within the replication of this classification system, research articles were specifically identified along with any research articles that were single-subject design or case studies.

3. RESULTS

The results of this exhaustive literature search on the topic of classroom-based AT used by students with visual impairments covered the 3.75-year time period from September 2009 to May 2013. We located 233 articles; however, 39% ($n=91$) were excluded because they did not meet inclusion criteria. Specifically, 30 articles did not involve any AT, 21 were published between January and August of 2009, 18 articles were not refereed publications (e.g., editorials or advertisements), 10 articles did not focus on AT for classroom-based educational interventions (e.g., orientation and mobility or independent living technology), 7 articles did not include participants with visual impairments, and 6 articles were medical studies on eye care health.

Consequently, there were 141 articles identified as published articles on AT that met the inclusion criteria for this analysis. Table 2.1 displays the numerical results of the current study and the numerical results of the original study by Kelly and Smith (2011). The information in Table 2.1 describes the current AT knowledge base as documented in peer-reviewed publications from September 2009 to May 2013 and addresses our first research question.

3.1. Characteristics of Participants in the Studies

At least 15% of the 141 articles that met the general criteria for the search included students who were visually impaired with additional disabilities. That is, at least 15% of all low vision and blindness-specific AT publications

Table 2.1 Characteristics of Published Assistive Technology Articles

Characteristic	Current Study		Original Study (2011)	
	n	%	*n*	%
Authors discussed theories, beliefs, or practices without a research design or method	38	27	122	48
Authors discussed product reviews or evaluation without a research design or method	23	16	34	13
Authors included a research design or method that did not include an intervention, control group, and comparison group **[a]Single-subject design was included in this characteristic ($N=5$)**	**13[a]**	**9**	**57[a]**	**22**
Authors presented an appropriate control and comparison group without sufficient resulting data to determine the effectiveness of the intervention within the manuscript	0	0	17	7
Participants were not students (e.g., parents, patients, families, or teachers)	47	33	11	4
Authors used a qualitative research design **[b]Case studies were included in this characteristic ($N=3$)**	**18[a]**	**13**	**6[b]**	**2**
Results and conclusions have validity issues (i.e., all participants received the treatment and the effects of the intervention could not be interpreted unambiguously)	0	0	4	2
A sighted comparison group was used inappropriately	0	0	3	1
Authors presented sufficient data to determine effectiveness of an intervention with appropriate participants, intervention, control group, and comparison group (i.e., scientifically based group research)	**2**	**1**	**2**	**1**
Total	141		256	

[a]Single-subject design was included in this category.
[b]Case studies were included in this category.
Note. Bold font type designates research articles discussed in subsequent tables.
Adapted with permission of AFB Press, American Foundation for the Blind, from Kelly and Smith (2011), p. 77;

included in this search from September 2009 to May 2013 included students who had additional disabilities. This percentage, however, is only an estimate based on the demographic information disclosed in each article. Articles that provided insufficient information about the disability status of the participants were not included in this estimate. For this reason, it is likely the number is higher than estimated. While this number of articles that represented students with additional disabilities may not be representative of the heterogeneous population of students with visual impairments, it was similar to the results of Kelly and Smith (2011) that included literatures from a 45-year time period. Kelly and Smith found that at least 20% of the articles published in a 45-year time period included students with visual impairments with additional disabilities.

In order to provide more information about the articles with a research basis, the 141 articles were reviewed to determine which articles used a methodology that included a research design with student participants and an AT intervention. The Council for Exceptional Children Division for Research (2004) conceptual groups (experimental and quasi-experimental group designs, single-subject designs, correlational designs, and qualitative designs) were used to subdivide the research-focused articles. Of the 141 articles, 33 studies (23%) met criteria for one of these four subcategories that included a methodology, the correct type of participants (students), and some type of AT-specific intervention.

3.2. Single-Subject Research

Approximately 38% ($n=5$; 15% of the 33 research articles) of the studies that were classified in Table 2.1 as including a research design or method without an intervention, control group, and comparison group were single-subject research studies. The single-subject research included a research design or method that demonstrated experimental control within a single participant or small group of participants (Kennedy, 2005). For low-incidence populations such as visual impairments, single-subject designed studies are a mechanism to complete experimental studies on students with visual impairments. The five studies used different types of single-subject designs, but all consistently focused on the impact of AT (see Table 2.2).

3.2.1 Math Assessment Using Calculator with Voice Input and Output

Bouck, Flanagan, Joshi, Sheikh, and Schleppenbach (2011) studied the efficiency of using a computer-based voice input, speech output (VISO) calculator. The study used a repeated acquisition design (see Kennedy, 2005)

Table 2.2 Summary of Single-Subject Design Studies ($N=5$)

Study	Participants w/VI	Intervention/Technology	Method/ Purpose	Content Area
Bickford and Falco (2012)	$N=9$ Ages 6–8 yrs Legally blind Braille readers	Patterns with Perkins Brailler or PAC Mate	SSD—alternating treatments	Braille literacy
Bouck et al. (2011)	$N=3$ Ages 18–19 yrs Blind and low vision	Voice input/speech output (VISO) calculator	SSD—alternating treatments	Math assessment
Kamei-Hannan and Lawson (2012)	$N=3$ Ages 12–14 yrs Braille users	Perkins Braille Writer, BrailleNote M-Power	SSD—alternating treatments	Braille literacy
Lancioni et al. (2010)	$N=2$ Ages 5–9 yrs Blind and low vision, severe disabilities	Microswitch clusters including sensors to detect student behavior and trigger access to preferred stimuli	SSD—withdrawal	Self-determination
Lusk (2012)	$N=5$ Grades 3–11 Low vision	Mounting devices for Color Pocket Viewer including handheld, stand-mounted, spectacle-mounted, and electronic	SSD—multielement design	Print reading

where students were asked to complete basic math problems using the VISO calculator on 20 assessments with 10 questions each. The study focused on efficiency (speed and number of attempts) as compared to the students' typical method of calculation (talking calculator). The authors found that initially students were more efficient with their traditional calculation method, but over time, performance with the VISO was equivalent. However, the students reported that the new technology had great potential as it provided them greater independence for solving complex problems.

3.2.2 Electronic Braille Notetakers and Literacy

Kamei-Hannan and Lawson (2012) studied the impact of using a braille notetaker versus a traditional Perkins Braille Writer on engagement and attitude during writing activities as well as quality of writing. Using an alternating-treatment design without a baseline with three students, the researchers found that students engaged in writing activities for longer periods and with higher quality when using the braille notetaker. They also found that the students were able to use the many features of the braille notetaker (such as spell check and cut/copy/past functions) that were not available with the Perkins Braille Writer. Overall, they surmised that braille notetakers were better tools for writing for these students and hypothesized that these findings may be generalizable to similar students. In a similar fashion, Bickford and Falco (2012) compared differences in braille fluency with instruction with traditional braille versus refreshable braille devices. Using an alternating-treatment design, they found that there were no significant differences in student performance on reading and writing braille. They found that students were motivated to use the new technology but that teachers were concerned about the ability of the students to learn spatial relationships when using the refreshable braille display only.

3.2.3 Microswitches and Self-Determination

Lancioni et al. (2010) conducted an A-B-A-B design experiment with two students with multiple disabilities (cognitive and physical) to determine if microswitches could be used to increase object-manipulation while reducing inappropriate behavior. They found that the use of microswitches that provided positive feedback had positive effects on both behaviors for these two students. The authors hypothesized that the use of devices such as microswitches allowed students to control their own environment with positive results on behavior and participation in activities.

3.2.4 Access to General Curriculum Through Magnifiers

Lusk (2012) used a multielement design to determine the optimal optical devices for students with low vision under five conditions. These conditions included the use of standard correction with appropriate print and four different types of magnifiers (handheld, stand-mounted, spectacle-mounted, and video). Lusk found that performance and preference varied across students for multiple reasons; she provided evidence that the selection of optimal magnification devices is based on individual needs and personal preferences.

3.3. Correlational Research

Of the 13 articles classified in Table 2.1 as including a research design or method without an intervention, control group, and comparison group, four used surveys or questionnaires and examined correlational relationships (see Table 2.3 for a complete listing). In several instances, authors used secondary analysis of large-scale nationally representative databases such as the National Longitudinal Transition Study-2 (NLTS2). Zhou, Griffin-Shirley, et al. (2012) assessed the relationship between computer Internet use and standardized test scores in secondary grade level students with visual impairments who participated in the NLTS2. Students who used a computer to access the Internet outperformed their peers who were not online on standardized test scores. Zhou, Griffin-Shirley, et al. (2012) discussed the importance of special assistance for students who were from low-income families based on their finding that these students reported a significantly lower rate of Internet use than students with visual impairments from high-income families (Zhou, Griffin-Shirley, et al., 2012).

Table 2.3 Summary of Correlational Studies ($N = 4$)

Study	Participants w/ VI	Intervention/ Technology	Method/ Purpose	Content Area
Freeland, Emerson, Curtis, and Fogarty (2010)	$N = 280$ Ages 13–17 yrs 33% Blind, 67% low vision	Use of access technology (type not specified)	Correlational—regression analysis of NLTS2 data	Standardized assessment
Kelly (2011)	$N = 1150$ Ages 13–16 yrs Low vision, blind, with and without severe disabilities	Use of access technology (type not specified)	Correlational—multilevel modeling of NLTS2 data	General
Kelly and Wolffe (2012)	$N = 520$ Ages 17–25 yrs Low vision and blind	Use of the Internet	Correlational—multilevel modeling of NLTS2 data	General
Zhou, Griffin-Shirley, et al. (2012), Zhou, Ajuwon, et al. (2012)	$N = 280$ Secondary students	Computer use for homework and Internet access	Correlational—regression analysis of NLTS2 data	Standardized assessment

The NLTS2 database was also used in correlational studies by Kelly (2011) and Kelly and Wolffe (2012). Kelly demonstrated that nearly 60% of high school students with visual impairments were not using AT and that changes in AT use over time were not significant. The statistically significant predictors of AT use were parent involvement, school placement, and likelihood of paid work experience (Kelly, 2011). Likewise, Kelly and Wolffe found that nearly 60% of the transition-aged youths with visual impairments were not using the Internet regularly for online communication and changes in online communication over time were not significant. The transition-aged youths who communicated online regularly had a significantly higher likelihood of being engaged in postsecondary work, education, training, or community service (Kelly & Wolffe, 2012).

3.4. Qualitative Research

As shown in Table 2.1, the single largest research category was qualitative research. Qualitative inquiry aims to gather in-depth understanding of behavior and determine the underlying possible reasons for the behavior (Shank, 2006). In this review, 18 articles (56%) were categorized as qualitative research, including multiple case studies. The range of topics and methodologies addressed in these 18 studies was very heterogeneous. Interventions included technology such as audio recordings, vision aids, videophones, screen readers, magnifiers, iPads, and many other high-tech to low-tech devices. The types of studies and the participants varied across the studies, thus each study must be reviewed independently to determine if the intervention may have potential for positive outcomes for students (see Table 2.4).

An example of this research includes an investigation of the barriers and challenges of AT use among students with visual impairments in Singapore (Wong & Cohen, 2011). Wong and Cohen (2011) found that the students with visual impairments included in their study and their teachers were not being adequately trained to use AT. For example, "Sunil, a 10 year old who uses Braille, is aware of the computer keyboard but is unable to touch type. When asked who will teach him to type he replied: 'Nobody. . .nobody will teach me'" (Wong & Cohen, 2011, p. 134). According to Wong and Cohen, there is an urgent need to better prepare teachers in the area of AT used by students with visual impairments through preservice and in-service teacher training. The qualitative research identified by this literature search provided rich details and vivid descriptions of the AT

Table 2.4 Summary of Qualitative Studies ($N=18$)

Study	Participants w/VI	Intervention/Technology	Method/Purpose	Content Area
Adetoro (2012)	$N=123$ Secondary school Blind and low vision	Braille, large print, or talking books/audio recordings	Qualitative (interview and questionnaire)—preference survey	Literacy—reading
Beal et al. (2011)	$N=14$ Grades 5–12 Mostly braille users	AnimalWatch-VI-Beta on Mac OSX	Qualitative (user data)—program evaluation, post hoc comparisons	Math instruction and assessment
Bengisu (2010)	$N=80$ Turkish adults 50% Congenitally blind, 50% low vision	Screen readers for cell phones (TALK, Mobile Speak), locators, laser cane, Arcelik appliances, reading pen, screen readers for computer (GVZ, Turkish Jaws)	Qualitative (phone survey)—product rating	Independent living skills (ECC)
D'Andrea (2012)	$N=12$ Ages 16–22 yrs Braille users	Various products for braille production, reading, listening, audio recording, screen reading, scanning, phone accessibility, and mobility	Qualitative (interviews)—product use and preferences	General
Emerson and Bishop (2012)	$N=10$ Ages 16–20 yrs Deafblind	Sorenson VP-200 videophone	Qualitative (phone logs, interviews, motivation rating scale)—user experience	Social networking (ECC)
Foley (2011)	$N=2$ College age Low vision	Screen reader and screen magnification software (not specified), user settings in Windows XP	Qualitative (interviews, questionnaire, and user data)—usability testing	Website accessibility

Freire, Linhalis, Bianchini, Fortest, and Pimentel (2010)	*N*=1 Academic student Blind	Interactive Whiteboard Software, with live mediator and accessible descriptions of graphics, accessed via keyboard and screen reading software (Virtual Vision)	Qualitative case study (interview and direct observation)—user experience	Geography
Hansen et al. (2010)	*N*=4 Ages 17–20 yrs 50% Low vision, 50% blind	Adaptive Content with Evidence-based Diagnosis (ACED) with talking tactile tablet for blind users	Qualitative (interviews and user data)—user experience	Math instruction and assessment
Helps and Herzberg (2013)	*N*=1 Age 17 yrs Low vision and multiple disabilities	iPad 2 with low-tech modifications (high contrast tape, nonstick mat, 3-ring binder)	Qualitative (direct observation)—task independence	Leisure (ECC)
Howard et al. (2012)	*N*=9 Middle school Severity of VI not specified	Lego Mindstorms NXT Robot Kit, JAWS, MAGIC magnifier, haptic feedback via Wii remote, and auditory feedback via PC	Qualitative (surveys and direct observation)	Robot programming
Klingenberg (2012)	*N*=2 Ages 10–11 yrs Blind	Pegboard box, raised line drawing kit, gym equipment	Qualitative (interview and direct observation)	Math instruction

Continued

Table 2.4 Summary of Qualitative Studies ($N = 18$)—cont'd

Study	Participants w/VI	Intervention/Technology	Method/Purpose	Content Area
Leung et al. (2010)	$N=1$ Age 11 yrs Low vision, physical disability	Character stroke disambiguation software with Big Buddy Button and laptop computer	Qualitative (user data)	Keyboarding for communication
Ludi and Reichlmayr (2011)	$N=46$ Grades 7–12 Blind, low vision	Lego Mindstorms NXT Robot Kit, JAWS, Zoomtext, BricxCC open-source environment	Qualitative (surveys, direct observation, discussion)—program evaluation	Robot programming
Siu and Lam (2012)	Not reported	Computer-assisted learning facilities	Qualitative (interviews, direct observation, product analysis)—program evaluation	Computers
Soderstrom and Ytterhus (2010)	$N=11$ Ages 15–18 yrs 55% Low vision, 45% blind	Information and communication technology	Qualitative (interviews)—user experience	General
Wong and Cohen (2011)	$N=11$ Ages 8–13 yrs 27% Low vision, 73% blind	High-tech AT including text-to-speech devices, screen readers, and screen magnifying software	Qualitative (interviews)—product use	General
Wormnaes and Sellaeg (2013)	$N=12$ Teachers Blind	Audio descriptions of video recordings	Qualitative (interviews and direct observations)—user experience	Listening comprehension
Yalo et al. (2012)	$N=78$ Teachers of students with VI in Grades 7–8, no students with VI	Low vision aids and teacher support for meeting needs of low vision students	Qualitative (interviews and direct observations)—identify challenges and solutions	General

experiences of students with visual impairments in particular contexts and geographic regions of the world. The qualitative research category also included three case studies (Freire et al., 2010; Helps & Herzberg, 2013; Klingenberg, 2012).

3.5. Descriptive, Quasi-Experimental, and Experimental Group Designs

Six of the 33 research studies (18%) used either a descriptive, quasi-experimental, or experimental design using AT (see Table 2.5). Descriptive research is used to describe characteristics of a population, but it cannot determine causality (Hinkle, Wiersma, & Jurs, 2003). There were two articles that fit this category (Brown et al., 2013; Zebahazy, Zigmond, & Zimmerman, 2012). Brown et al. is an example of descriptive research that assessed perspectives on the classroom environment experienced by students with visual impairments. They used widely available questionnaires that had been previously validated by others (e.g., the *Teacher Opinion Questionnaire* and the *Parent Involvement Questionnaire*) to examine classroom environments in the regular Australian early education system and found inadequacies in how students with visual impairments were served within these systems.

As reported by several previous authors in the United States (Abner & Lahm, 2002; Corn & Wall, 2002; Edwards & Lewis, 1998; Kapperman, Sticken, & Heinze, 2002; Parker et al., 1990), teachers reported that they received little to no training with the AT required by the students with visual impairments in their classrooms or on their caseloads. Students with visual impairments did not have access to the required AT due to both teacher knowledge deficits and limited financial resources for equipment purchases (Brown et al., 2013). This absence of AT probably constrained the learning and development of students with visual impairments in early childhood education according to Brown et al. (2013). These examples of survey research show the wide application and broad coverage of the survey technique (i.e., descriptive research) despite not having an intervention, control group, or comparison group.

There were only two studies identified by the analysis that included a research design with sufficient data to determine the effectiveness of an intervention with an intervention, control group, and comparison group (Kapperman et al., 2011, 2012). The latter study (Kapperman et al., 2012) was a replication, follow-up, and continuation of the original study by Kapperman et al. (2011) that evaluated the long-term effectiveness of the intervention. In both instances, the studies evaluated the effectiveness

Table 2.5 Summary of Experimental, Quasi-Experimental, and Descriptive Group Designs ($N=6$)

Study	Participants w/VI	Intervention/ Technology	Method/Purpose	Content Area
Beal and Shaw (2009)	$N=11$ Middle school Blind	PC with JAWS, supplemented with braille and abacus	Descriptive (observational research)—pre/post	Math assessment
Brown, Packer, and Passmore (2013)	$N=20$ Kindergarten–Grade 1	Vision aids and equipment	Descriptive(survey and direct observation)—availability of resources, post hoc comparisons	General
Boffoli et al. (2011)	$N=12$ Ages 9–16 yrs Low vision	Exergames including Dance Dance Revolution Extreme 2, Eye Toy Kinetic, and Wii Boxing	Randomized group design—no control	Leisure
Kapperman, Sticken, and Smith (2011)[a]	$N=57$ Grades 5–11 Braille users	Nemeth code tutorial for BrailleNote	Randomized group design with control group	Math assessment
Kapperman, Sticken, and Smith (2012)[a]	$N=41$ Grades 6–12 Braille users	Nemeth code tutorial for BrailleNote	Follow-up to Kapperman et al. (2011)—pre/post study of control group	Math assessment
Zebehazy et al. (2012)	$N=286$ Grades 3, 4, 7, and 8 Visual learners, nonvisual learners	Accommodations such as slant board, alternate communication forms, pictures, objects	Descriptive—post hoc analysis of group differences based on severity of VI	Alternate assessment

[a]*Note*: The two Kapperman et al. studies (2011, 2012) are the only experimental studies with appropriate participants, interventions, control group, and comparison groups. The other studies are either quasi-experimental or descriptive studies.

of a tutorial installed on a BrailleNote for learning the Nemeth Code of braille math notation (Kapperman et al., 2011, 2012).

The authors reported using an experimental design in which a sample of braille-reading students who completed the software tutorial on the BrailleNote ($n = 28$) was matched with a control group of braille-reading students ($n = 28$) according to their grade, math achievement, and ethnicity (Kapperman et al., 2011, 2012). Student performance on reading and writing mathematical symbols using Nemeth braille math notation was measured across participants. Kapperman et al. (2011, 2012) showed that treatment groups had significantly greater growth in both math reading and math writing than the control group. This finding persisted one year after treatment, and long-term effectiveness of the BrailleNote Nemeth Code tutorial was evident (Kapperman et al., 2012). Thus, the two quasi-experimental studies identified by this exhaustive review of the literature document the effectiveness of AT designed specifically to help braille students learn Nemeth math symbols.

Therefore, regarding our second research question, our field has made minimal progress in determining the effectiveness of AT for individuals with visual impairments. A substantial number of articles have been published in the area of classroom-based AT used by students who are visually impaired. However, the use of experimental group research has not progressed during the time period assessed by the present study.

4. DISCUSSION

The purpose of the original literature review by Kelly and Smith (2011) was to identify evidence-based practices (or "promising practices") for AT for the education of individuals with visual impairments. That study, intended to be a meta-analysis, became a literature synthesis due to the lack of experimental research, included 256 articles from 1965 to August 2009. During that 44-year period, only two studies were found that presented sufficient data to determine the effectiveness of an intervention with appropriate participants, intervention, control group, and comparison group. These criteria were consistent with the selection criteria used by the WWC (What Works Clearinghouse, 2011). Consequently, Kelly and Smith (2011) presented a challenge to the field to conduct more scientifically sound research on the educational use of AT with students with visual impairments.

This present study set out to replicate the previous study but went beyond the original methodology to expand and discuss the major types of research and findings. This research study covers 3.75 years (September 2009 to May 2013) during which 141 articles were published that met the initial criteria for inclusion. A further step was taken in this study by categorizing the "research" articles as defined by CEC-DR (2004). Comparing the two studies provides one obvious finding; the amount of literature on the topic of AT has increased greatly. In the Kelly and Smith (2011) study, which spanned 44.75 years or 536 months, an average of 5.72 articles were published per year (or 0.48 articles per month). Of course, this basic summary is a crude measure as most of the technology articles were written in the latter years of the study. In comparison, this current study spanned 3.75 years, or 45 months. During that much shorter time period, an average of 37.6 articles per year, or 3.13 articles per month, were published. While this is a very rudimentary comparison of the quantity of literature on AT in our field, one thing is clear—the number of peer-reviewed publications dealing with AT in special education has increased dramatically in the past few years.

Closer analysis of the literature revealed similarities and differences between the two studies. For example, this study's literature had almost half as many studies on theory, beliefs, and practices (27% compared to 48% in 2011). There was an increase of studies with participants who were individuals other than students with visual impairments (33% compared to 4% in 2011). There was a substantial increase in qualitative research articles (13% compared to 2% in 2011) with a decrease in studies that did not include an intervention, control group, and comparison group (9% compared to 22%). Ironically, the two studies both only had two published studies that met the rigorous standards.

4.1. Challenges to the Research

4.1.1 Low Incidence of Participants

While it is encouraging that there are more peer-reviewed publications on AT for educational purposes, the limited research on the impact of AT is an important issue. While the criteria used by the WWC are scientifically valid, the ability to conduct "gold standard" randomized experimental research with large numbers of participants is incredibly challenging for the field of visual impairments (Ferrell, 2007). With the majority of students with visual impairments included in general education and smaller pools of students at residential schools for the blind, recruiting and randomly assigning the large samples needed for confidence in statistical procedures limit our research options. The uniqueness of individual students, the heterogeneity of the impact of specific eye diseases and disorders on learning abilities, and the general lack of research funding make it easy to understand the lack of rigorous scientific research on this population (Ferrell, 2007). Therefore, research on the use of AT with students with visual impairments must use different methods, such as single-subject designs that are now accepted by the WWC.

4.1.2 Diversity of AT Resources

Another challenge is the range of assistive technologies that are available to meet the needs of individuals with visual impairments. Because the population of students with visual impairments is so diverse, a large array of assistive technologies is required to meet their needs. For example, the American Foundation for the Blind (AFB, 2012) maintains an online database of assistive technologies for individuals with visual impairments categorized into 30 areas with a large number of products for each area. AFB lists 106 different products under "CCTV/Video Magnifiers," 29 products under "Refreshable Braille Displays," and 20 products under "Educational Technology." The disability-specific AT used by people with visual impairments can be expensive, leading some consumers to shift to more universally designed technology such as the iPad and other popular iOS devices. These off-the-shelf assistive technologies evolve each year as technology becomes more sophisticated.

4.1.3 Students are not Using Technology as Much as Peers

At the same time, there is also evidence that students with visual impairments are not using technology at the same rates as their sighted peers. Kelly and Smith (2008) found that 77% of preadolescents with visual impairments did not use social networking (email and/or chat rooms), dramatically lower than their nondisabled peers (Rideout, Foehr, & Roberts, 2010). In a 2009 study, Kelly found that only 29–41% of students with visual impairments who were most likely to benefit from the use of AT were actually using it during the three waves of data collected for the Special Education Elementary Longitudinal Study. Kelly also found that the use of technology did not increase over the 5-year period (2000–2004) and that parental involvement and educational placement significantly impacted the use of AT. Kelly (2009) were consistent with the correlational findings of Kelly (2011) and Kelly and Wolffe (2012).

Results from these studies suggest that many students with visual impairments are not using technology to the extent possible leading us to question "why"? Many barriers may contribute to their lack of AT use including cultural attitudes or bias against technology (Carey, DelSordo, & Goldman, 2004), personal attitudes (Zascavage & Keefe, 2004), lack of resources (Lahm, 2003; Lee & Vega, 2005), and lack of instructional time (Collier, Weinburgh, & Rivera, 2002). However, the literature reveals that the most prominent barrier to AT use may be the lack of technology skills among qualified professional personnel, specifically teachers of students with visual impairments (Abner & Lahm, 2002; Candela, 2003; Lee & Vega, 2005; Zascavage & Keefe, 2004).

4.1.4 Teacher Competency with AT

Multiple researchers have documented that teachers of students with visual impairments report that they did not feel competent enough to teach their students to use assistive technologies (Abner & Lahm, 2002; Edwards & Lewis, 1998; Kapperman et al., 2002; Zhou, Ajuwon, et al., 2012; Zhou, Parker, Smith, & Griffin-Shirley, 2011; Zhou, Smith, Parker, & Griffin-Shirley, 2011). In a recent national survey of 840 professionals who work with students with visual impairments, Zhou, Ajuwon, et al. (2012) found that 59.3% of the participants reported "no" to "some" confidence in teaching and supporting the use of AT. Direct instruction from qualified professionals within the field of visual impairment is required to achieve competency with the sophisticated AT used by people who are visually impaired (Kapperman & Sticken, 2000) and to access commonly used and universally designed devices. While this is not the only barrier to students' effective use of AT, it is definitely an important one that should be amenable to intervention. Many factors may influence the lack of competence reported by professionals who work with students with visual impairments. One aspect may be the lack of instruction at the preservice level. Smith and Kelley (2007) found that only half of the personnel preparation programs in the United States and Canada reported having a specific course that focused on AT. The other universities included AT instruction as part of other courses, special workshops, or even embedded instruction within the entire program (Pogrund & Smith, 2012).

An issue at the heart of the lack of instruction at the preservice (and possibly at the in-service level also) may be the lack of competencies/standards for teachers that focused on AT. The initial licensure requirements for TVIs from the Council for Exceptional Children (CEC, 2003) have limited items related to AT. Competencies from other leading organizations (International Society of Technology in Education and Rehabilitation

Engineering and Assistive Technology Society of North America) have very limited standards related to the unique set of competencies needed for educators who work with students with visual impairments. Therefore, Smith and his colleagues (Smith, Kelley, Maushak, Griffin-Shirley, & Lan, 2009) completed a study using the Delphi method to establish a set of AT competencies for preservice teachers of students with visual impairments along with specific levels of expertise. These researchers identified 111 competencies with accompanying level of expertise. The competencies were later validated by practicing TVIs at the state (Zhou, Parker, et al., 2011; Zhou, Smith, et al., 2011) and national levels (Zhou, Ajuwon, et al., 2012). The field now has an established set of competencies to guide pre- and in-service preparation in AT for students with visual impairments.

4.1.5 Rapidly Evolving AT

While the lack of instruction of AT is a major barrier for students with visual impairments, there is another issue that has an impact on both students and their teachers. As already noted, technology is a driving force in the society, and it is constantly evolving to become faster, smaller, and more efficient. This rapid change in technology creates a unique challenge for researchers to test the effectiveness of educational and assistive technologies and for practitioners to become expert instructors of the technology. It is commonly accepted that AT has positive impacts on the lives and opportunities of individuals with disabilities, including individuals with visual impairments (Cooper & Nichols, 2007; Kapperman et al., 2002; Strobel, Fossa, Arthanat, & Brace, 2006). However, this belief may be rooted in the general idea that advanced technology supersedes all other practices and may not necessarily be based on scientific evidence. For example, it is generally accepted that the Perkins Brailler is "better" than the slate and stylus. While it may be true that the Perkins Brailler is faster and more efficient for preparing braille assignments, stating that it is "better" is not rooted in research. In some situations or contexts such as writing labels or writing quick notes when shopping, the slate and stylus may actually outperform the Perkins Brailler. All of these challenges have had a major impact on the level of intervention-based research on assistive technologies for students with visual impairments.

However, many of the articles on AT for students with visual impairments do have merit and inform practice. While the WWC does not review qualitative studies, they provide information about the complex issues related to motivation and underlying reasons for phenomenon. Qualitative research in AT focuses on developing a set of "informed assertions" that can drive other research (Shank, 2006). As noted, single-subject designed studies provide a more "rigorous, scientific methodology used to define basic principles of behavior and establish evidence based practices" (Horner et al., 2005, p. 165). Single-subject research can document causal, or functional, relationships between independent variables (interventions) and dependent variables for socially important outcomes (Horner et al., 2005). Due to its level of rigor, the What Works Clearinghouse (2010) established a mechanism for conducting single-subject study "reviews" as part of their work (Kratochwill et al., 2010). Qualitative, descriptive, and single-subject designed research can provide insight into "promising practices" and help guide future quantitative research.

The relationships among the different types of research must be respected and used to the fullest extent. Research methods should be based on the questions being asked; with so little research in our field, we may need to use a variety of methods to provide the foundation for the rigorous and expensive randomized controlled trials that are required by the WWC. Even though research on the educational impact of AT may be mixed, the opinions of individuals with visual impairments clearly indicate that they feel that AT is critical to their personal success (D'Andrea, 2012). The social validity of its use is reason enough for students with visual impairments to receive instruction on the use of a variety of assistive technologies.

4.2. Limitations

Although this was a replication study of Kelly and Smith (2011), advancements or changes in the field since the original study was conceptualized required slight modifications to our procedure to be consistent with the original conceptual framework. There were some differences in electronic databases that were available for the present study. For example, EBSCO was replaced with Proquest Educational Journals as a platform to access the ERIC and Psychinfo databases. After assessing the databases currently available and comparing their range with each of the databases available for Kelly and Smith, we adopted the five electronic databases that corresponded most precisely with of the electronic search we implemented for the earlier study. Also, the search terms list was expanded to included technology that has been developed recently. For example, the terms "refreshable braille display," "app," "interactive whiteboards," and "haptic touch" were added to the Boolean search technique. The manual search procedures also required some minor adjustments. The manual search procedures included only relevant journals that were in print during the time period assessed by the present study. This resulted in the inclusion of new journals in the manual search and exclusion of journals that were no longer in print during the time period assessed by the present study.

4.3. Future Research

In light of limited research, researchers in the field of special education of students with visual impairments need to consider novel steps to improve the body of research on AT in education. A first step would be to establish a strong research agenda that formalizes and focuses the work of the field. A research agenda would establish the major questions and themes that need to be answered and the priority in which they should be addressed. For instance, research on single input (audio or tactile/braille) versus multiple input (combinations of both audio and tactile) is a major question that relates to multiple AT devices. Developing a research agenda can be an arduous task but the literature may already provide insight into aspects of AT that may need to be addressed. As mentioned earlier, Smith et al. (2009) used the Delphi method, an iterative process of reaching consensus on a particular subject using the knowledge of an expert panel, to identify AT competencies to guide preservice instruction and in-service professional development. The exhaustive list of 111 AT competencies might be used to guide future research (Smith, 2008).

For example, Competency 12 (C12), "The effects of low vision on the use of AT, such as lighting, contrast, size, glare" (Smith et al., 2009, p. 462), could lead to research that tests the effectiveness and efficiency of specific assistive technologies for students with low vision in educational settings. Another example can be found in C13 that states, "braille literacy and its application in providing effective assistive technology" (Smith et al., 2009, p. 462). Research on the effectiveness of AT in teaching braille literacy could provide the foundation for educational decision making. One of the domains is entitled "Use of Assistive Technology" and encompasses 22 competencies that list specific technologies that all could be researched for effectiveness and/or efficiency (Smith et al., 2009, p. 463).

Even with the research agenda established, there are larger issues that the field must address to advance research on AT. These issues include scarcity of qualified researchers, lack of research funding, difficulty of locating large numbers of participants, and ever evolving technology. However, these issues could be circumvented by a focused effort by the entire field. For instance, the establishment of a science, technology, engineering, and mathematics, or STEM, consortium composed of the leading researchers in the field could (1) establish a research community, (2) create a focused research agenda, (3) create collaborative partnerships, (4) share resources, (5) establish large-scale agreements with AT developers and vendors, and (6) pool geographically diverse participants into single studies.

Due to the challenges in conducting rigorous research on AT with individuals with visual impairments, the first step of the STEM consortium would be to bring the community together in a research community a group of colleagues engaged in a collaborative effort to support the grand themes of research within the field in question. The purpose of this consortium would be to formally establish an association that not only conducts collaborative research but that also hosts meetings, workshops, symposiums, and/or conferences.

A major benefit of such a consortium would be the possibility of engaging the AT industry as partners in the research. Funding for research for students with visual impairments has long been underfunded (Corn & Ferrell, 2000; Mason, Davidson, & McNerney, 2000). A collaborative effort, with the commitment of the industry, would provide opportunities for larger grant opportunities, possibly even the creation of a research center. But beyond funding opportunities, having the relationship with the AT industry would allow researchers to support the development of technologies while they are actually being developed. Such a relationship would benefit everyone as researchers would have access to the latest technologies while providing expert insight to developers with all focusing on providing students with assistive technologies with a solid research base. Lastly, the STEM consortium could establish formal relationships with consumer organizations of the blind such as the National Federation of the Blind and the American Council of the Blind. These organizations already have established efforts in these areas but might benefit from formal connections to more researchers in the field.

5. CONCLUSION

This study set out to determine the "state of the knowledge base" in AT for students with visual impairments. Using the rigorous standards set in the Kelly and Smith (2011) literature synthesis, which is based loosely on standards set by the WWC, we found that the amount of rigorous, scientifically based research has not increased since the original study. However, we did find an increase in the number of peer-reviewed publications on the topic of AT for education of students between 3 and 21 years of age. As the field evolves, we predict that the research base will continue to not only grow but also become more rigorous. For more desperately needed evidenced-based practices to emerge, the field will need to drastically change the way in which research on AT is conducted through a focused research agenda and collaboration across fields and industries to ensure that there is sufficient evidence to support the use of specific technologies.

REFERENCES

Abner, G., & Lahm, E. (2002). Implementation of assistive technology with students who are visually impaired: Teachers' readiness. *Journal of Visual Impairment & Blindness, 96*, 98–105.

Adetoro, N. (2012). Alternative format preferences among secondary school visually impaired students in Nigeria. *Journal of Librarianship and Information Science, 44*(2), 90–96.

American Foundation for the Blind. (2012). Products database. http://www.afb.org/aw/main.asp.

Attewell, P., & Battle, J. (1999). Home computers and school performance. *The Information Society, 15*, 1–10.

Beal, C. R., & Shaw, E. (2009). An online math problem solving system for middle school students who are blind. *Journal of Online Learning and Teaching, 5*(4), 1–9.

Beal, C. R., Rosenblum, L., & Smith, D. W. (2011). A pilot study of a self-voicing computer program for prealgebra math problems. *Journal of Visual Impairments & Blindness, 105*(3), 157–169.

Bengisu, M. (2010). Assistive technologies for visually impaired individuals in Turkey. *Assistive Technology, 22*(3), 163–171.

Bickford, J. O., & Falco, J. R. (2012). Technology for early braille literacy: Comparison of traditional braille instruction and instruction with an electronic notetaker. *Journal of Visual Impairment & Blindness, 107*, 679–693.

Boffoli, N., Foley, J. T., Gasperetti, B., Yang, S. P., & Lieberman, L. (2011). Enjoyment levels of youth with visual impairments playing different exergames. *Insight: Research and Practice in Visual Impairment and Blindness, 4*(4), 171–176.

Bouck, E. C., Flanagan, S., Joshi, G. S., Sheikh, W., & Schleppenbach, D. (2011). Speaking math-A voice input, speech output calculator for students with visual impairments. *Journal of Special Education Technology, 26*(4), 5–14.

Brown, C. M., Packer, T. L., & Passmore, A. (2013). Adequacy of the regular early education classroom environment for students with visual impairment. *Journal of Special Education, 46*, 223–232.

Candela, A. R. (2003). A pilot course in teaching skills for assistive technology specialists. *Journal of Visual Impairment & Blindness, 97*, 661–666.

Carey, A. C., DelSordo, V., & Goldman, A. (2004). Assistive technology for all: Access to alternative financing for minority populations. *Journal of Disability Policy Studies, 14*, 194–203.

Collier, S. A., Weinburgh, M. H., & Rivera, M. (2002). Infusing technology skills into a teacher education program: Change in students' knowledge about and use of technology. *Journal of Technology and Teacher Education, 12*, 447–468.

Cooper, H. L., & Nichols, S. K. (2007). Technology and early braille literacy: Using the Mountbatten Pro Brailler in primary-grade classrooms. *Journal of Visual Impairment & Blindness, 101*, 22–31.

Corn, A. L., & Ferrell, K. A. (2000). External funding for training and research in visual disabilities at colleges and universities, 1997-98. *Journal of Visual Impairment & Blindness, 94*, 372–384.

Corn, A. L., & Lusk, K. E. (2010). Perspectives on low vision. In A. L. Corn & J. N. Erin (Eds.), *Foundations of low vision: Clinical and functional perspectives* (pp. 3–34) (2nd ed.). New York: AFB Press.

Corn, A., & Wall, R. (2002). Access to multimedia presentations for students with visual impairments. *Journal of Visual Impairment & Blindness, 96*, 197–211.

Council for Exceptional Children. (2003). *What every special educator must know: Ethics, standards, and guidelines for special educators.* Arlington, VA: In press.

Council for Exceptional Children Division for Research. (2004). Quality indicators for research in special education and guidelines for evidence-based practices: Executive summary. http://www.cecdr.org/pdf/QI_Exec_Summary.pdf.

D'Andrea, F. M. (2012). Preferences and practices among students who read braille and use assistive technology. *Journal of Visual Impairment & Blindness, 106*, 585–596.

Edwards, B. J., & Lewis, S. (1998). The use of technology in programs for students with visual impairments in Florida. *Journal of Visual Impairment & Blindness, 92*, 302–312.

Emerson, J., & Bishop, J. (2012). Videophone technology and students with deaf-blindness: A method for increasing access and communication. *Journal of Visual Impairments & Blindness, 106*(10), 622–633.

Ferrell, K. (2007). *Issues in the field of blindness and low vision.* National Center on Severe and Sensory Disabilities. http://www.unco.edu/ncssd/resources/issues_bvi.pdf.

Foley, A. (2011). Exploring the design, development and use of websites through accessibility and usability studies. *Journal of Educational Multimedia and Hypermedia, 20*(4), 361–385.

Freeland, A. L., Emerson, R., Curtis, A. B., & Fogarty, K. (2010). Exploring the relationship between access technology and standardized test scores for youths with visual impairments: Secondary analysis of the National Longitudinal Transition Study 2. *Journal of Visual Impairment & Blindness, 104*, 170–182.

Freire, A. P., Linhalis, F., Bianchini, S. L., Fortest, R., & Pimentel, M. (2010). Revealing the whiteboard to blind students: An inclusive approach to provide mediation in synchronous e-learning activities. *Computers & Education, 54*, 866–876.

Hansen, E. G., Shute, V. J., & Landau, S. (2010). An assessment-for-learning system in mathematics for individuals with visual impairments. *Journal of Visual Impairments & Blindness, 104*(5), 275–286.

Hatton, D. D., Ivy, S. E., & Boyer, C. (2013). Severe visual impairments in infants and toddlers in the United States. *Journal of Visual Impairment & Blindness, 107*(5), 325–336.

Helps, D. H., & Herzberg, T. S. (2013). The use of an iPad2 as a leisure activity for a student with multiple disabilities. *Journal of Visual Impairment & Blindness, 107*, 232–236.

Hinkle, D. E., Wiersma, W., & Jurs, S. G. (2003). *Applied statistics for the behavioral sciences.* Boston: Houghton Mifflin.

Horner, R. H., Carr, E. G., Halle, J., McGee, G., Odom, S., & Wolery, M. (2005). The use of single-subject research to identify evidence-based practice in special education. *Exceptional Children, 71*, 165–179.

Howard, A. M., Park, C. H., & Remy, S. (2012). Using haptic and auditory interaction tools to engage students with visual impairments in robot programming activities. *Journal of Visual Impairments & Blindness, 5*(1), 87–95.

Hunley, S. A., Evans, J. H., Delgado-Hachey, M., Krise, J., Rich, T., & Schell, C. (2005). Adolescent computer use and academic achievement. *Adolescence, 40*, 307–318.

Individuals with Disabilities Education Improvement Act. (2004). 20 U.S.C. § 1400 *et seq*.
Jackson, L. A., von Eye, A., Biocca, F. A., Barbatsis, G., Zhao, Y., & Fitzgerald, H. E. (2006). Does home Internet use influence the academic performance of low-income children? *Developmental Psychology*, *42*, 429–435.
Johnson, S. S., & Evans, J. (2005). Considering responses efficiency as a strategy to prevent assistive technology abandonment. *Journal of Special Education Technology*, *20*, 55–57.
Judge, S., Puckett, K., & Bell, S. (2006). Closing the digital divide: Update from the Early Childhood Longitudinal Study. *Journal of Educational Research*, *100*, 52–60.
Kamei-Hannan, C., & Lawson, H. (2012). Impact of a Braille-Note on Writing: Evaluating the process, quality, and attitudes of three students who are visually impaired. *Journal of Special Education Technology*, *27*, 1–14.
Kapperman, G., & Sticken, J. (2000). Assistive technology. In In A. J. Koenig & M. C. Holbrook (Eds.), *Foundations of education. Vol. II.* (pp. 500–528) (2nd ed.). New York: AFB Press.
Kapperman, G., Sticken, J., & Heinze, T. (2002). Survey of the use of assistive technology by Illinois students who are visually impaired. *Journal of Visual Impairment & Blindness*, *96*, 106–108.
Kapperman, G., Sticken, J., & Smith, T. J. (2011). The effectiveness of the Nemeth Code Tutorial for the BrailleNote. *Insight: Research and Practice in Visual Impairment and Blindness*, *4*, 22–28.
Kapperman, G., Sticken, J., & Smith, T. J. (2012). The effectiveness of the Nemeth Code Tutorial for the BrailleNote: A follow-up study. *Insight: Research and Practice in Visual Impairment and Blindness*, *5*, 168–174.
Kelly, S. M. (2009). Use of assistive technology by students with visual impairments: Findings from a national survey. *Journal of Visual Impairment & Blindness*, *103*, 470–480.
Kelly, S. M. (2011). Assistive technology use by high school students with visual impairments: A second look at the current problem. *Journal of Visual Impairment & Blindness*, *105*, 235–239.
Kelly, S. M., & Smith, T. J. (2008). The digital social interactions of students with visual impairments: Findings from two national surveys. *Journal of Visual Impairment & Blindness*, *102*, 528–539.
Kelly, S. M., & Smith, D. W. (2011). The impact of assistive technology on the educational performance of students with visual impairments: A synthesis of the research. *Journal of Visual Impairment & Blindness*, *105*, 73–83.
Kelly, S. M., & Wolffe, K. (2012). Internet use by transition-aged youth with visual impairments in the United States: Assessing the impact of postsecondary predictors. *Journal of Visual Impairment & Blindness*, *106*, 597–608.
Kennedy, C. H. (2005). *Single-case design for educational research*. Boston: Allyn and Bacon.
Klingenberg, O. G. (2012). Conceptual understanding of shape and space by braille-reading Norwegian students in elementary school. *Journal of Visual Impairment & Blindness*, *106*, 453–465.
Koenig, A. J., & Ashcroft, S. C. (1983). Initial assessment of student use of the electric Perkins Brailler. *Education of the Visually Handicapped*, *15*, 59–66.
Kratochwill, T. R., Hitchcock, J., Horner, R. H., Levin, J. R., Odom, S. L., Rindskopf, D. M., et al. (2010). *Single-case designs technical documentation*. Retrieved fromWhat Works Clearinghouse. website: http://ies.ed.gov/ncee/wwc/pdf/wwc_scd.pdf.
LaGrow, S. J. (1981). Effects of training on CCTV reading rates of visually impaired students. *Journal of Visual Impairment & Blindness*, *75*, 368–373.
Lahm, E. A. (2003). Assistive technology specialists: Bringing knowledge of assistive technology to school districts. *Remedial and Special Education*, *24*, 141–153.

Lancioni, G. E., O'Reilly, M. F., Singh, N. N., Sigafoos, J., Didden, R., Oliva, D., et al. (2010). Two children with multiple disabilities increase adaptive objective manipulation and reduce inappropriate behavior via a technology-assisted program. *Journal of Visual Impairment & Blindness*, *104*, 714–719.

Lee, Y., & Vega, L. A. (2005). Perceived knowledge, attitudes, and challenges of AT use in special education. *Journal of Special Education Technology*, *20*, 60–62.

Leung, B., Yates, M., Duez, P., & Chau, T. (2010). Text entry via character stroke disambiguation for an adolescent with severe motor impairment and cortical visual impairment. *Assistive Technology*, *22*, 223–235.

Lowenfeld, B. (1973). *The visually handicapped child in school.* New York: John Day.

Ludi, S., & Reichlmayr, T. (2011). The use of robotics to promote computer to pre-college students with visual impairments. *ACM Transactions on Computing Education*, *11*(3), 20: 1–20:20.

Lusk, K. E. (2012). The effects of various mounting systems of near magnification on reading performance and preferences in school-age students with low vision. *British Journal of Visual Impairment*, *30*, 168–181.

Mason, C., Davidson, R., & McNerney, C. (2000). *National plan for training personnel to serve children with blindness and low vision.* Reston, VA: The Council for Exceptional Children.

Michaels, C. A., & McDermott, J. (2003). Assistive technology integration in special education teacher preparation: Program coordinators' perceptions of current attainment and importance. *Journal of Special Education Technology*, *18*, 29–41.

No Child Left Behind (NCLB) Act of 2001. (2002). Pub. L. No. 107-110, § 115, Stat. 1425.

Parker, S., Buckley, W., Truesdell, A., Riggio, M., Collins, M., & Boardman, B. (1990). Barriers to the use of assistive technology with children: A survey. *Journal of Visual Impairment & Blindness*, *84*, 532–533.

Peterson-Karlan, G. R., & Parette, H. P. (2007). Evidence-based practice and the consideration of assistive technology: Effectiveness and outcomes. *Assistive Technology Outcomes and Benefits*, *4*, 130–139.

Pogrund, R., & Smith, D. W. (2012). Impact of a short-term training model on assistive technology for preservice teachers of students with visual impairments. *Insight: Research and Practice in Visual Impairment & Blindness*, *5*, 100–110.

Rideout, V., Foehr, U. G., & Roberts, D. F. (2010). *Generation M2 Media in the lives of 8- to 18-year-olds (Report 8010).* Menlo Park, CA: Henry J. Kaiser Family Foundation.

Sacks, S. Z. (1998). Educating students who have visual impairments with other disabilities: An overview. In S. Z. Sacks & R. K. Silberman (Eds.), *Educating students who have visual impairments with other disabilities* (pp. 3–38). Baltimore, MD: Paul H. Brookes Publishing.

Schafer, W. D. (2001). Replication: A design principle for field research. *Practical Assessment, Research & Evaluation*, 7. http://PAREonline.net/getvn.asp?v=7&n=15.

Shank, G. D. (2006). *Qualitative research: A personal skills approach* (2nd ed.). Upper Saddle River, NJ: Pearson.

Siu, K. W. M., & Lam, M. S. (2012). Public computer assisted learning facilities for children with visual impairment: Universal design for inclusive learning. *Early Childhood Education Journal*, *40*, 295–303.

Smith, D. W. (2008). *Assistive technology competencies for teachers of students with visual impairments: A Delphi Study.* Unpublished doctoral dissertation. Lubbock, TX: Texas Tech University.

Smith, D. W., & Kelley, P. A. (2007). A survey of the integration of assistive technology knowledge into teacher preparation programs for individuals with visual impairments. *Journal of Visual Impairments & Blindness*, *101*, 429–433.

Smith, D. W., Kelley, P. A., Maushak, N., Griffin-Shirley, N., & Lan, W. Y. (2009). Assistive technology competencies for teachers of students with visual impairments: A Delphi study. *Journal of Visual Impairments & Blindness, 103*, 457–469.
Soderstrom, S., & Ytterhus, B. (2010). The use and non-use of assistive technologies from the world of information and communication technology by visually impaired young people: a walk on the tightrope of peer inclusion. *Disability & Society, 25*(3), 303–315.
Strobel, W., Fossa, J., Arthanat, S., & Brace, J. (2006). Technology for access to text and graphics for people with visual impairments and blindness in vocational settings. *Journal of Vocational Rehabilitation, 24*, 87–95.
What Works Clearinghouse. (2010). *Single-case design technical documentation.* http://ies.ed.gov/ncee/wwc/documentsum.aspx?sid=229.
What Works Clearinghouse. (2011). *Procedures and standards handbook (Version 2.1).* http://ies.ed.gov/ncee/wwc/pdf/reference_resources/wwc_procedures_v2_1_standards_handbook.pdf.
Wittwer, J., & Senkbeil, M. (2008). Is students' computer use at home related to their mathematical performance at school? *Computers & Education, 50*, 1558–1571.
Wong, M. E., & Cohen, L. (2011). School, family and other influences on assistive technology use: Access and challenges for students with visual impairments in Singapore. *British Journal of Visual Impairment, 29*, 130–144.
Wormnaes, S., & Sellaeg, N. (2013). Audio-described educational materials: Ugandan teachers' experiences. *British Journal of Visual Impairment, 31*(2), 164–171.
Yalo, J. A., Indoshi, F. C., & Agak, J. O. (2012). Challenges and strategies of working with learners with low vision: Implications for teacher training. *Educational Research and Review*, 7(10), 238–243.
Zascavage, V. T., & Keefe, C. H. (2004). Students with severe speech and physical impairments: Opportunity barriers to literacy. *Focus on Autism and Other Developmental Disabilities, 19*, 223–234.
Zebahazy, K. T., Zigmond, N., & Zimmerman, G. J. (2012). Performance measurement and accommodation: Students with visual impairments on Pennsylvania's Alternate Assessment. *Journal of Visual Impairment & Blindness, 106*, 17–30.
Zhou, L., Ajuwon, P. M., Smith, D. W., Griffin-Shirley, N., Parker, A. T., & Okungu, P. (2012). Assistive technology competencies for teachers of students with visual impairments: A national study. *Journal of Visual Impairment & Blindness, 106*, 656–665.
Zhou, L., Griffin-Shirley, N., Kelley, P., Banda, D. R., Lan, W. Y., Parker, A. T., et al. (2012). The relationship between computer and internet use and performance on standardized tests by secondary students with visual impairments. *Journal of Visual Impairment & Blindness, 106*, 609–621.
Zhou, L., Parker, A. T., Smith, D. W., & Griffin-Shirley, N. (2011). Assistive technology for students with visual impairments: Challenges and needs in teachers' preparation programs and practice. *Journal of Visual Impairment & Blindness, 105*, 197–210.
Zhou, L., Smith, D. W., Parker, A. T., & Griffin-Shirley, N. (2011). Assistive technology competencies of teachers of students with visual impairments: A comparison of perceptions. *Journal of Visual Impairments & Blindness, 105*, 533–547.

CHAPTER THREE

Teaching Skill Acquisition to Individuals with Blindness: A Systematic Review of Response-Prompting Procedures

Sarah E. Ivy[1], Deborah D. Hatton
Department of Special Education, Vanderbilt University, Nashville, Tennessee, USA
[1]Corresponding author: e-mail address: sarah.e.ivy@vanderbilt.edu

Contents

Abstract

A search of the literature on prompting procedures to teach skill acquisition yielded 22 single-subject studies that included participants with little to no functional vision. Studies were published between 1984 and 2012 in 13 different journals. Quality

International Review of Research in Developmental Disabilities, Volume 46
ISSN 2211-6095
http://dx.doi.org/10.1016/B978-0-12-420039-5.00005-8

indicators for single-subject research (Horner et al., 2005) were applied to identify studies that met high quality standards. Findings from high-quality studies were aggregated to determine if prompting could be considered an evidence-based practice (EBP) for this population. Across the 12 studies meeting high quality standards, 8 different research groups taught 35 different behaviors to 27 participants. Behaviors included chained and discrete skills across domains of communication, motor skills, self-care, vocational/leisure, and community/living. System of least prompts was used most frequently with several procedural variations. Research supports the conclusion that prompting is an evidence-based practice to teach new skills to learners with blindness.

1. INTRODUCTION

Learners with visual impairment (VI) constitute a small but exceptionally diverse group in terms of ability level and visual functioning. It is estimated that 1 in 1000 school-age children and youth meets eligibility standards for VI and 10% of students with VI are functionally blind and cannot learn through vision (Erin, 2003). Due to the significant overlap between categories of disability used to determine eligibility of students to receive special education services, data from the U.S. Department of Education underestimate the prevalence of students with VI. Because students are categorized according to their eligibility, estimating the number of students with VI who have additional disabilities is also complicated. Students with VI who have additional disabilities may be counted under the category of multiple disabilities. According to the U.S. Department of Commerce, Economics and Statistics Administration, U.S. Census Bureau (2011), one-third of school-age children with VI not attending residential schools for the blind in 2010 had a cognitive disability. It has also been estimated that 65% of students with VI have multiple disabilities (Parker & Pogrund, 2009). Recently, a national registry of infants and toddlers with VI reported that 68% of young children with severe, uncorrectable VI referred to specialized agencies also had multiple disabilities (Hatton, Ivy, & Boyer, 2013).

A visual impairment can have profound effects on many areas of development (Hatton, Bailey, Burchinal, & Ferrell, 1997; Warren & Hatton, 2003). For example, early in a child's life, caregivers often find it difficult to establish a healthy attachment with their children who do not make eye contact (Realon, Favell, & Dayvault, 1988). Additional social consequences begin to emerge when children start attending school and fail to pick up on nonverbal social cues (Preisler, 1993). Also children are less motivated to explore their environment because their curiosity is not being stimulated by their visual surroundings (Sonksen, Levitt, & Kitzinger, 1984). It has been shown repeatedly that children reach for sound later than the visually directed reach is developed (e.g., Brambring, 2007). This lack of exploration affects motor development as well as concept development (Bruce & Muhammad, 2009; Fraiberg, 1977). Unfortunately, children with VI fall further behind in conceptual, skill, and social development due to the lack of incidental learning that typically takes place from observing others (Bigelow, 1995). Finally, even when direct instruction is provided, many children with VI have difficulty maintaining and generalizing skills because reinforcement is often in visual form. Compound the developmental implications of having a visual impairment with additional disabilities, and it is not difficult to understand the struggle to develop language and communication, social, and functional skills to achieve academically and have a high quality of life.

Historically, this group of students has been among the children neglected by the educational system (Wolery & Gast, 2000). However, provided with adequate supports, students with multiple disabilities and VI have learned functional skills and basic communication (Ferrell, 2006; Parker, Davidson, & Banda, 2007; Parker & Pogrund, 2009). Often, this support has come in the form of response-prompting procedures, a category of instructional strategies from applied behavior analysis, which transformed education for students with significant disabilities (Wolery & Schuster, 1997).

1.1. Response-Prompting Instructional Strategies

Prompting strategies are used when students do not readily acquire naturally reinforcing behaviors. Because the behaviors rarely occur, they are rarely reinforced. Prompting strategies are used to increase the occurrence of a target behavior and in doing so increase the likelihood the behavior will be reinforced. Reinforcement increases the probability the behavior will reoccur given the same stimulus conditions in the future. Response prompting includes providing physical assistance, providing verbal instructions, and/or modeling behavior to elicit a target behavior. Stimulus control refers to the prompt or cue responsible for eliciting behavior. It is essential to transfer stimulus control from supporting prompts to naturally occurring stimuli in the environment for students to achieve independence. Wolery and Gast (1984) described four methods for fading response prompts: system of least prompts, most-to-least prompting, graduated guidance (GG), and time delay. Additionally, simultaneous prompting (SP) has been increasingly used and is recognized as an effective and systematic type of response prompting (Waugh, Alberto, & Fredrick, 2011). Each of these response-prompting procedures is included in the subject of this chapter.

1.1.1 Time Delay

Time delay was first introduced by Touchette (1971) to fade stimulus prompts, and the term was later coined by Striefel, Wetherby, and Karlan (1978). It has since been adapted to fade response prompts and recently applied to teach spelling (e.g., Coleman-Martin & Heller, 2004), sight word reading (e.g., Collins, Evans, Creech-Galloway, Karl, & Miller, 2007; Reichow & Wolery, 2011), decoding (e.g., Cohen, Heller, Alberto, & Fredrick, 2008), purposeful communication (e.g., Dipipi, Jitendra, & Miller, 2001), picture or object identification (e.g., Dogan & Tekin-Iftar, 2002; Tekin-Iftar, Acar, & Kurt, 2003), switch use for social communication (e.g., Cosbey & Johnston, 2006), sports and leisure skills (e.g., Tekin-Iftar et al., 2001), numeral identification (e.g., Akmanoglu-Uludag & Batu, 2005), color recognition (e.g., Çiftçi & Temel, 2010), recognition of product warning labels (e.g., Dogoe, Banda, Lock, & Feinstein, 2011), food preparation (e.g., Bozkurt & Gursel, 2005; Fiscus, Morse, Schuster, & Collins, 2002), purchasing skills (e.g., DiPipi-Hoy & Jitendra, 2004), and aquatic play skills (e.g., Yilmaz, Birkan, Konukman, & Erkan, 2005).

Primarily, time delay procedures vary by the amount of delay inserted between the stimulus and the prompt. Two types of time delay procedures are constant time delay (CTD) and progressive time delay (PTD). CTD is characterized by a time delay (usually 5 s) that is held constant across all opportunities to respond. A CTD procedure typically follows a sequence whereby (1) the stimulus is presented (e.g., in the form of a question), (2) the controlling prompt is delivered after a 0 s delay (e.g., the answer), (3) the stimulus is presented in subsequent trials, and (4) the controlling prompt is delivered after a 5 s delay in all subsequent trials. PTD is similar except the time delay is gradually increased with subsequent opportunities to respond by some standard increment (usually 0.5, 1, or 2 s). SP is not generally referred to as a time delay strategy because the prompt is always delivered concurrently with the stimulus (Waugh et al., 2011); however, it is like time delay in which the delay is always 0 s. Opportunities to respond without prompts are inserted on a regular basis to assess whether the behavior is under stimulus control.

Time delay or SP may be the first choice of prompting strategy to teach students new targets (e.g., recognize new sight words) within a class of behaviors they already have in their repertoire (e.g., ability to read print). Common inclusion criteria in time delay studies include the ability to wait, respond to simple task instructions, and perceive the target stimulus; however, procedures for wait training and providing attending cues have been described in the literature and shown to be effective (Walker, 2008; Wolery et al., 1992). Overall, simple skills requiring a single behavioral response (i.e., discrete skills) are more common than skills requiring multiple behavioral responses (i.e., chained skills) as targets for time delay instruction. Preschoolers and elementary students are more often taught discrete skills, while the majority of participants in studies using time delay to teach chained tasks have been adolescents (Schuster et al., 1998; Walker, 2008; Wolery et al., 1992). Time delay and simultaneous prompts have been used most often with students who have mild to moderate intellectual disabilities and infrequently with students who have severe or profound intellectual disabilities (Browder, Ahlgrim-Delzell, Spooner, Mims, & Baker, 2009; Waugh et al., 2011; Wolery et al., 1992). However, researchers who have used time delay to teach chained skills, as compared to discrete skills, have included participants with severe or profound intellectual disabilities more often (Schuster et al., 1998). In studies of time delay to teach chained skills to participants with severe disabilities, the number of 0 s delay trials extended to as many as four sessions (Dogoe & Banda, 2009), compared to a single session or less in studies of time delay to teach discrete skills to participants with mild to moderate disabilities (Wolery et al., 1992). However, the delay interval was comparable across these sets of studies (i.e., on average 4 s).

In terms of efficiency, PTD may require less instructional time to reach mastery than CTD, but CTD may result in learning that maintains longer (Schuster et al., 1998; Wolery et al., 1992). Additionally, CTD is a simpler procedure, which may make training professional staff easier and lead to higher fidelity of implementation (Schuster et al., 1998). Few studies have compared SP to CTD, and mixed results suggest the procedures are fairly comparable in effectiveness and efficiency with SP perhaps being slightly more efficient (Schuster et al., 1998; Wolery et al., 1992).

1.1.2 Prompting Hierarchies

Prompting hierarchies call for a change in the type of prompt delivered on subsequent opportunities to respond. The earliest use of prompting hierarchies in research was devised to teach janitorial skills to adults with intellectual disabilities (Brown, Bellamy, Perlmutter, Sackowitz, & Sontag, 1972). Prompting hierarchies vary in the levels of prompts that are planned, but three is a common number (e.g., physical assistance, modeling, and verbal). The different types of prompting hierarchies are named according to their directionality.

A hierarchy of prompts that begins with the least intrusive prompt (e.g., verbal instruction) and ends with the most intrusive prompt (e.g., full physical assistance) characterizes the system of least prompts (SLP). A SLP sequence for teaching a child to turn on the water faucet might include (1) allowing the child the opportunity to perform the skill independently, (2) telling the child to "turn it on," (3) demonstrating how to turn it on, and finally (4) providing hand-over-hand or hand-under-hand assistance. Usually, a specified amount of wait time is planned before the prompt is administered at each level. SLP has been used effectively in recent years to teach use of voice output communication devices (e.g., Dyches, Davis, Lucido, & Young, 2002; Sigafoos, O'Reilly, Seely-York, & Edrisinha, 2004), use of office equipment and ATMs (Taber-Doughty, 2005), household chores (West & Billingley, 2005), laundry skills (Taylor, Collins, & Schuster, 2002), and object communication in the context of story book reading (Mims, Browder, Baker, Lee, & Spooner, 2009). In a review of studies that used SLP, 62% of skills taught across studies were chained tasks; however, the youngest participants were always taught discrete skills. Adolescents with severe or profound intellectual disability were also frequently taught discrete communication or language skills. Most adults, despite their level of intellectual functioning, were taught chained skills applicable to community settings (Doyle, Wolery, Halle, & Gast, 1988).

Most-to-least prompts (MLP) are similar to SLP except the procedure begins with the most intrusive prompt and moves down the hierarchy with subsequent opportunities to respond until independence is reached. With the MLP system, several trials are performed at each level until the intrusiveness of the prompt is decreased. Among other skills, MLP systems have been used effectively to teach toy use (Libby, Weiss, Bancroft, & Ahearn, 2008), switch use for toy play (Realon et al., 1988), and banking skills (McDonnell & Ferguson, 1989).

Azrin and Armstrong first described GG in 1976. Since then, the defining characteristics of the procedure have gradually changed. For this procedure, a controlling prompt is selected, and then, the amount of assistance is reduced over time as the student begins to perform an activity independently. The difference between GG and MLP or SLP is that interventionists implementing GG have greater flexibility to decide the level of prompting to administer based on student responses during the activity (Neitzel & Wolery, 2009). Previously, the controlling prompt was always a full or partial physical prompt and changed only in the level of intensity or position of the prompt (e.g., wrist, elbow, or shoulder; Demchak, 1990; Wolery & Gast, 1984). Recently, the definition has expanded to include the use of different types of prompts (i.e., not just physical) for the controlling prompt and to fade the controlling prompt (Neitzel & Wolery, 2009). Often, an opportunity to respond independently is provided before the controlling prompt is administered (Neitzel & Wolery, 2009). Among other skills, GG has been used to teach self-feeding (Luiselli, 1993) and leisure skills (Demchak, 1990). According to Wolery and Gast (1984), MLP and GG have most often been used to teach chained gross motor skills.

MLP is often used to teach skill acquisition for behaviors not already in a student's repertoire, whereas SLP may be a better option to increase fluency of behaviors already occurring (Demchak, 1990). In the case of teaching a new skill, Wolery and Gast (1984) recommended the use of SLP to determine the level of prompt required for a student to perform a behavior with 100% accuracy and then the use of most-to-least prompting to fade support. In terms of efficiency, because children are given the opportunity to respond independently before prompts are administered with an SLP procedure, there exists more room for student error. Frequency of errors to mastery criterion is a common way researchers measure the efficiency of prompting procedures; therefore, it could be assumed that MLP would be more efficient than SLP on this measure. On other measures of efficiency, it is unclear if one prompting hierarchy is the most efficient.

1.1.3 Choosing Types of Prompts

When choosing the types of prompts to use, it may be important to consider the reinforcing nature of physical prompts for students. Physical prompts may negatively reinforce behavior for students who are tactually sensitive. Alternatively, waiting behavior may be positively reinforced with students who enjoy physical contact, which may lead to prompt dependency. These may be important factors to consider when defining procedures for use with students who are blind because tactual sensitivity is common for this population (Spencer, Head, Van Dusen Pysh, & Chalfant, 1997). If a decreasing hierarchy of prompts is desirable, Demchak (1990) suggested that a GG procedure might be more effective than MLP for students who are averse to physical guidance because opportunities to respond independently are more frequent and occur sooner. Another important consideration may be if and how to include modeling within a system of prompts. Often modeling is a type of prompt used in SLP procedures, which may be why Wolery and Gast (1984) suggested that SLP should be reserved for students who imitate. Because initiation is typically visually motivated, and most children who are blind do not imitate visual models, modeling may take the form of hand-under-hand guidance. Therefore, imitation behavior may be an inappropriate inclusion or exclusion criteria for students with VI. On the other hand, because students with VI often struggle to develop imitation and initiation, instructors may select an MLP over SLP.

Time delay strategies are frequently cited as more efficient (i.e., requiring fewer trials/sessions to criterion, less instructional time, and fewer errors to criterion) than prompting hierarchies to teach discrete and chained skills (Demchak, 1990; Handen & Zane, 1987; Schuster et al., 1998; Wolery et al., 1992). Five studies reviewed by Schuster et al. (1998) compared CTD and SLP to teach chained skills and found that CTD was more efficient in all cases. Two studies compared CTD and MLP and found that CTD was more efficient to teach laundry skills and MLP was more efficient to teach banking skills. Walker (2008) reviewed two studies comparing PTD and SLP and found PTD to be more efficient to teach object and sight word identification in terms of sessions to criterion and/or errors to criterion. Time delay strategies and SP are also more conducive to group instruction than are prompting hierarchies. In a review of 47 studies using response prompting in small-group instruction, time delay strategies were used in all but one study, which used SLP (Ledford, Lane, Elam, & Wolery, 2012). Group instruction is desirable to increase opportunities for incidental learning through observation; however, often, the materials and modes of instruction are not accessible to students with VI.

1.1.4 Summary

Prompting strategies are systematic yet highly flexible instructional tools. They have been effective to teach a diverse group of learners a diverse group of skills. Important considerations for selecting the "right" prompting strategy include the nature of the target skill and characteristics of the learner including age, preferences, and the presence and severity of a sensory impairment or intellectual disability. Time delay strategies appear to be more commonly used to teach discrete skills to learners with mild or moderate disabilities. Prompting hierarchies appear to be more commonly used to teach chained skills to learners with more severe disabilities. Empirical evidence suggests that time delay strategies are more efficient than prompting hierarchies in general; however, comparison data between MLP and CTD are inconclusive. At this time, we do not have sufficient empirical evidence to predict the effect of VI on learners' success with different prompting strategies.

1.2. Efficacy of Response-Prompting Strategies

For a practice to be considered evidence-based, we must specify for whom and under what conditions it is effective (Kratochwill et al., 2010). What constitutes sufficient research is somewhat arbitrary, and the field is still in the process of defining evidence-based practices (EBP; Cook & Odom, 2013). However, Horner et al. (2005) suggested that five or more high-quality single-subject research studies, conducted by at least three different research groups, that demonstrate the efficacy of an independent variable with a minimum of 20 participants should be sufficient evidence.

Response prompting is an area of research that has a long and rich history. Numerous published studies have demonstrated the effectiveness and relative efficiency of these strategies to teach a range of skills, given particular sets of learning variables (Demchak, 1990; Dogoe & Banda, 2009; Morse & Schuster, 2004; Schuster et al., 1998; Walker, 2008; Waugh et al., 2011; Wolery et al., 1992). The Horner et al. (2005) quality indicators have been used to identify response prompting to teach discrete skills during small-group instruction (Ledford et al., 2012) and time delay to teach literacy skills (Browder et al., 2009) as EBPs for students with disabilities. Prompting strategies, as a whole, have been identified as an EBP for students with autism (National Professional Center on Autism Spectrum Disorders, 2010).

1.3. Evidence-Based Practices for Students with VI

An evidence-based instructional strategy has not yet been identified for students with visual impairment (Ferrell, 2006; Parker & Pogrund, 2009). A small number of the studies included in reviews on response prompting included participants with low vision (Bennett, Gast, & Schuster, 1986; Coleman-Martin & Heller, 2004; Hamre-Nietupski, Nietupski, Sandvig, & Ayres, 1984; Venn et al., 1993). However, in each of these studies, some aspect of the prompting system was visual. Additionally, in their review of prompting within the context of small-group instruction, Ledford et al. (2012) determined that the most common inclusion criterion was visual acuity good enough to see target stimulus and prompts. Because generality of findings using single-subject methodology is based on inclusion criteria, results from these studies are generalizable to students who can use vision to complete tasks. Unfortunately, because students with blindness are not among the groups represented in reviews on response prompting, the results cannot be generalized to these students. Therefore, despite the thorough review of response-prompting procedures, very little is known about the extent to which prompting is effective for students with blindness.

According to Horner and colleagues (2005), it is prudent to begin the search for EBPs among interventions that are already widely thought to be effective. Response prompting has been operationally defined and is already widely used in classrooms serving students with visual impairment. Also, there is a large body of literature on prompting; increasing the likelihood that sufficient research has already been published to identify it as an evidence-based practice for students with VI.

1.4. Research Questions

In this chapter, we synthesize the findings from research on response-prompting procedures to answer the following questions: (a) What types of prompting strategies are being used to teach what skills under what conditions? (b) What are the characteristics of participants with little to no functional vision (i.e., blindness) for whom response prompting has been used to teach skill acquisition? (c) Is response prompting an evidence-based practice for teaching functional skills to students with blindness? Specifically, do the single-subject research studies that have been published meet the Council for Exceptional Children's Division on Research criteria (CEC-DR) for high-quality research, and does the body of research meet CEC-DR criteria for an EBP across studies (Horner et al., 2005)?

2. METHOD

We used four search methods to identify articles for this chapter. First, a database search resulted in an initial list of intervention articles. We derived an additional list of potential intervention articles from a hand search of the premier journal for educational research on visual impairment in the United States, *Journal of Visual Impairment & Blindness* (*JVIB*). A third list of potential intervention articles came from an ancestral search of the intervention articles included in the final chapter. Finally, another database search for literature reviews on prompting was conducted, and those articles were reviewed for additional intervention articles. The following are the details of each search method, inclusion and exclusion criteria, and a description of the methods used to evaluate the intervention articles included in this chapter.

2.1. Search Terms and Method

An electronic search of titles and abstracts, using the PsycINFO, ERIC, and PsycARTICLES databases, was conducted combining the terms "vis* impair*," "vis* loss," "vis* disab*," "multipl* disab*," "deafblind*," "deaf-blind*," "deaf blind," "dual* impair*," "sensory impair*," "multi-handicap*," and "vis* handicap*" with "prompt*," "graduated guidance," and "time delay." The search was restricted to articles from peer-reviewed journals published in English. No date restrictions were used. With duplicates deleted, this search yielded 73 articles. For the first pass, the first author reviewed only the abstracts of each article. Potential intervention articles were excluded if one or more of the following were clearly indicated: (a) A prompting procedure was not used, (b) the intervention relied on visual supports such as pictures or visual modeling, or (c) a single-case design was not used. Studies with multiple levels of independent variables were included as long as student descriptions and data were reported. Using this procedure, 41 potential intervention studies were found.

For the second pass, the first author reviewed the full text of the intervention studies to apply the inclusion and exclusion criteria. Of the 41 studies, 37 included students with visual impairment. However, 10 of the 37 were excluded because either all participants had good residual vision or their vision was not described in enough detail to determine the severity. Another nine were excluded for not using a systematic prompting procedure, and two were excluded because researchers used prompting to increase a rate of behavior rather than teach a new skill. The remaining 16 studies were included in this chapter.

In addition to the database search, the first author conducted a manual search of articles published in *JVIB* since 1994. The year 1994 was used as a cutoff date because *JVIB* published a review of applied behavior analysis strategies used with students with VI that year (DeMario & Crowley, 1994). First, only the titles were examined to extract the references of articles that might meet the final inclusion criteria. Then, the abstract and/or full text was retrieved for each reference extracted, and the full inclusion and exclusion criteria were applied. No additional studies meeting the inclusion criteria were discovered using this method.

Next, the first author searched for review articles using ERIC, PsycINFO, and PsycARTICLES for article titles that contained the words, "prompt*," "time delay," or "graduated guidance" combined with "review" in the title or abstract. This search returned 12 reviews on the subject of prompting strategies. The 1994 *JVIB* review was also checked for additional intervention studies. Four additional studies were located using this method. Finally, the first author conducted an ancestral search of the intervention articles selected for inclusion in this chapter, which resulted in one additional study. One additional study was discovered through personal communication.

2.2. Inclusion and Exclusion Criteria

Intervention studies selected for this review targeted skill acquisition as the dependent variable and used (a) at least one school-age participant with little to no functional vision, (b) a response-prompting procedure from one of the categories described by Wolery and Gast (1984) or SP, and (c) a single-case experimental design. Studies using an A–B design were excluded. Intervention packages that included materials not present in baseline or other instructional strategies besides prompting and reinforcement were also excluded. Finally, participants were excluded whose visual impairment was not described in sufficient detail to determine the severity of the vision loss. Labels such as deafblind or legally blind without a functional description of residual vision were not sufficient for inclusion in this chapter; however, participants who were described as blind were included. If it could reasonably be determined that participants primarily interacted with the world through touch and/or sound or if a description of their vision indicated a minimal response to light and/or objects, participants were included in this chapter. Having an "inconsistent" visual response was not sufficient to determine participants' level of functional vision.

2.3. Interrater Agreement

For article selection, the second rater independently coded 20% of articles returned from the database search and 20% of issues published by *JVIB* since 1994. Interrater agreement was calculated using the point-by-point method by study for study inclusion and by quality indicator per study for applying the Horner et al. criteria. The formula for calculating agreement was the number of agreements divided by the number of agreements plus disagreements, multiplied by 100 (Ayres & Gast, 2010). Interrater agreement for article selection was 100%. The second rater coded 20% of included studies to apply the quality indicators and agreement was 96%. Two disagreements concerned the quality indicator for the description of the procedural parameters. These were discussed, and the conservative judgment was made to mark those indicators as not met.

2.4. Coding and Evaluation of Selected Studies

The evaluation of each study was based on the criteria for high-quality single-subject research from the *Council for Exceptional Children's Quality Indicators for Single-Subject Research* (Horner et al., 2005). Horner et al. identified 21 quality indicators to evaluate single-subject studies. We coded descriptive information pertaining to each quality indicator in detail, except for those quality indicators pertaining to social validity and external validity. We reasoned that although objective measures of social validity, maintenance, and generalization are highly desirable, they are not essential to establishing experimental control to demonstrate that an intervention can be effective. Thus, they were not included in our evaluation of the quality of this set of studies. The remaining quality indicators were consolidated into the following 11 criteria statements that were coded as either met or not met:

- Description of participants and inclusion criteria: Name of disability and assessment used for diagnosis or functional description given for all participants.
- Description of the setting and interventionist: Adequate information to determine the teacher–student ratio, location, and the relationship between the student and the interventionist.
- Description and measurement of the dependent variable: Construct was operationally defined in terms of observable behavior, and metric was reported and consistent from baseline to intervention.
- Description of procedures: Materials, amount of intervention, and procedural parameters (including baseline procedures) described in enough detail to replicate.
- Reliability of the dependent variable: Interobserver agreement collected for a minimum of 20% of sessions and reported at or above 80%, or kappa at or above 0.60.
- Fidelity of implementation: Fidelity collected objectively for at least 20% of sessions and reported above 90%.
- Independent variable was actively manipulated.
- Repeated measurement of behavior: At least three data points in each condition with data collected immediately prior to and after introduction of the independent variable.

- Experimental design: Experimental design used with at least three planned replications.
- Replication of effects: Functional relation demonstrated by three replications of effect.
- No attrition of participants.

Standards published by Horner et al. (2005) were also used to determine whether response-prompting procedures are an EBP for students who are blind, which are (a) five or more high-quality single-subject studies, (b) conducted by at least three different research groups, and (c) replicated with 20 participants.

The criteria statements listed previously were organized in three categories (i.e., description of study elements, reliability, and internal validity) and used to determine whether a study met high quality standards with or without reservations using the following system. To meet the highest quality standards (without reservations), all criteria had to be met (see Table 3.1 for categorization). To meet standards with reservations, all criteria for experimental control/internal validity (items 7–11) and the quality indicator for reliability of the dependent variable (item 5) had to be met. If these standards were not met, the studies were not included in the summative evaluation to determine if the published research on response prompting was sufficient to consider prompting an EBP for students with little to no functional vision.

If a study met standards with or without reservations, we used the method outlined by Ledford et al. (2012) to determine if the authors represented independent research groups. Specifically, studies with any overlapping authors were considered to be from the same research group.

Finally, because of the small number of studies that have used prompting procedures to teach students with severe VI, all studies with at least one student with a visual impairment that met the inclusion criteria were included in this chapter. Although all of the participants in each study were used to determine whether there was a functional relation, only the participants with VI were used to determine the effectiveness of the intervention with this group of learners. Hence, a functional relation was necessary but not sufficient to demonstrate effectiveness. A functional relation must be shown as well as the effect demonstrated by the student(s) with VI.

Visual analysis was used to determine if a change in the dependent variable occurred across conditions. Despite the recent push to calculate effect sizes for single-subject research, a suitable statistic to address the serial dependency of the data has not been identified (Cook & Cook, 2013; Odom, Dunst, Horner, & Bradley, 2012). Likewise, variations to calculating percentage of non-overlapping data do not yield consistent results (Wolery, Busick, Reichow, & Barton, 2010). Therefore, visual analysis was used to code each instance of change across conditions as a (a) change in level, (b) change in trend, (c) change in level and trend, or (d) no change. Additionally, descriptive information was coded for each study to allow detailed reporting regarding (a) participants' ages, diagnoses, severity of intellectual disability, and amount of functional vision; (b) target behaviors, settings, materials, and relationship between the interventionist and the participant; (c) procedural parameters of the prompting interventions and reinforcement procedures; and (d) outcome measures.

3. RESULTS

After inclusion and exclusion criteria were applied, 22 studies were included in this chapter. Studies were published from 1984 to 2012 and all were conducted in the United States. Out of 57 participants, 43 had a severe visual impairment that met the inclusion criteria. Each response-prompting procedure was used at least once; however SLP, MLP, or GG was used to teach 83.3% of the behaviors taught to participants across studies. In three studies, researchers referred to their procedures as GG; however, we found their procedures more precisely fit the Wolery and Gast (1984) definition for SLP (Correa et al., 1984) and MLP (McKelvey et al., 1992; Sisson et al., 1988). For this chapter, we categorized the interventions according to the Wolery et al. definitions.

Before quality indicators were applied to identify the studies meeting standards, a total of 90 replications were successful, across 60 skills and 43 participants with blindness. To test the efficacy of response prompting, a multiple baseline or multiple probe design was used in every study but one, which used a withdrawal design (Strawbridge et al., 1989). Studies were published in 13 different journals, primarily those that publish behavior- or vision-specific literature. *Journal of Visual Impairment & Blindness* (*JVIB*) published the largest number of studies on this topic; however, three of the four published in *JVIB* did not meet quality standards.

3.1. Prompting Strategies, Target Skills, and Contexts

Hierarchies and time delay were used to teach chained and discrete skills. SLP was used most often and was used to teach a range of skills to participants with a range of ability levels. Except for Wright and Wolery (2012), who used GG to teach street crossing, only SLP was used to teach participants with mild or no

Table 3.1 Extent to Which Studies Met Single-Subject Quality Indicators

Quality Indicators (Horner et al., 2005)	Wright and Wolery (2012)	Browder, Lee, and Mims (2011)	Mims et al. (2009)	Lancioni et al. (2008)	Lancioni et al. (2007)	Grisham-Brown, Schuster, Hemmeter, and Collins (2000)	Lancioni, O'Reilly, Campodonico, and Serenelli (2000)	Lancioni, O'Reilly, Oliva, Bianchi, and Pirani (2000)	Gee, Graham, Sailor, and Goetz (1995)	Trask-Tyler, Grossi, and Heward (1994)
Description of Study Elements										
1. Description of participants and inclusion criteria	•	•	•	•	•	•	•	•	•	•
2. Description of the setting and interventionist	•	•	•	•	•	•	–	–	•	•
3. Description and measurement of the dependent variable	•	•	•	•	•	–	•	•	•	•
4. Description of procedures	•	•	•	•	•	•	•	•	•	–
Reliability										
5. Reliability of the dependent variable	•	•	•	–	•	•	•	–	•	•
6. Fidelity of implementation	•	•	•	–	–	•	–	–	–	–
Experimental Control/Internal Validity										
7. Independent variable actively manipulated	•	•	•	•	•	•	•	•	•	•
8. Repeated measurement of behavior	•	•	•	•	•	•	•	•	•	•
9. Experimental design	•	•	•	–	•	•	–	–	•	•
10. Replication of effect	•	•	•	–	•	•	–	–	•	•
11. No attrition of participants	•	•	•	•	•	•	•	•	•	•

Authors' names are shaded for studies meeting quality standards for an evidence based practice as described in the methods section. Circles represent indicators met and dashes represent indicators not met.

Taylor, McKelvey, and Sisson (1993)	Taras, Matson, and Felps (1993)	McKelvey, Sisson, Van Hasselt, and Hersen (1992)	Collins, Gast, Wolery, Holcombe, and Leatherby (1991)	Reese and Snell (1991)	Gee, Graham, Goetz, Oshima, and Yoshioka (1991)	Berg and Wacker (1989)	Strawbridge, Drnach, Sisson, and Van Hasselt (1989)	Sisson, Kitwein, and Van Hasselt (1988)	Taylor (1987)	Sisson and Dixon (1986)	Correa, Poulson, and Salzberg (1984)
•	•	•	•	•	•	•	•	•	•	•	•
•	–	–	•	•	•	–	•	•	–	•	•
•	–	•	•	•	•	–	•	•	–	•	•
•	–	–	•	•	•	–	•	•	–	•	•
•	•	–	•	–	•	–	•	•	–	•	•
–	–	–	•	–	•	–	–	–	–	–	–
•	•	•	•	•	•	•	•	•	•	•	•
•	•	–	•	•	•	–	•	•	–	•	•
–	•	•	•	•	•	•	•	•	•	–	•
–	•	–	–	•	•	•	•	•	•	–	•
•	•	•	•	•	•	•	•	•	•	•	•

ID. For students with more significant ID, a wider a range of prompting strategies and top-down hierarchies were used. However, time delay strategies in particular were used in only a few instances with mixed results.

3.1.1 Target Skills

See Table 3.2 for a breakdown of skill category by participants' level of intellectual disability. In order of frequency from highest to lowest, target skills fell in to the following categories: community/living (29.0%, e.g., street crossing), self-care (26.9%, e.g., self-dressing), motor skills (19.4%, e.g., reach and grasp), language/communication (12.9%, e.g., requesting continuation of a routine), social/leisure (8.6%, e.g., renting a movie), and vocational (3.2%, e.g., stuffing envelopes). Almost exclusively, community/living skills were taught to children with mild or no intellectual disability, vocational and leisure skills taught to participants with mild to moderate

Table 3.2 Number of Inter- and Intrasubject Replications By Skill Domain and Participant Level of Intellectual Disability

		Participant's Level of Intellectual Disability				
Skill Domain		**None**	**Mild**	**Moderate**	**Severe**	**Profound**
Motor	Discrete	–	–	–	11[a–c]	9[a,d]
	Chained	–	–	–	–	–
Communication	Discrete	–	–	–	8[a–c]	2[e]
	Chained	–	–	–	–	–
Self-care	Discrete	–	–	–	2[a]	–
	Chained	–	–	7[b,d]	13[b–d]	2[a]
Leisure	Discrete	–	–	–	–	–
	Chained	–	4[a]	–	–	–
Vocational	Discrete	–	–	–	–	–
	Chained	–	3[a]	3[a]	–	–
Community/living	Discrete	–	–	–	–	–
	Chained	8[a,d]	15[a]	–	–	3[a]

[a]System of least prompts.
[b]Most-to-least prompting.
[c]Constant time delay.
[d]Graduated guidance.
[e]Progressive time delay.
Note: Self-care includes dressing and feeding. Living includes preparing foods.

ID, self-care skills taught to participants with moderate to profound ID, communication skills taught to children with severe ID, and motor skills taught to children with severe to profound ID. Skills were more often chained (62.4%) than discrete (37.6%). Discrete skills fell into the motor, communication, or self-care domain and were only taught to participants with severe to profound disabilities.

3.1.2 Procedural Parameters

Details about the procedural parameters for each type of prompting procedure are provided in Tables 3.3–3.6. In every study, baseline consisted of the same structured setting as during intervention, without prompting and in many cases without reinforcement. Verbal or tactile praise was provided as contingent reinforcement during intervention conditions in all but two studies (Strawbridge et al., 1989; Trask-Tyler et al., 1994). In the majority of studies (70%), contingent reinforcement procedures were different between baseline and intervention. In two exceptions, baseline conditions consisted of using reinforcement procedures to change behavior that would otherwise prevent skill acquisition. Sisson and Dixon (1986) reinforced sitting and eating without engaging in self-injurious behavior during baseline. After 24 sessions to decrease self-injurious behavior, intervention for napkin use began, followed by intervention for utensil use. Similarly, Strawbridge and colleagues (1989) used contingent reinforcement to increase the frequency that the student held the handles of his walker. Subsequently, reinforcement was held constant (music) while teaching stepping with the walker.

3.1.2.1 System of Least Prompts

SLP was the most often studied procedure, used in 14 of the studies in this sample. However, procedural parameters for SLP are perhaps the most homogenous of the response-prompting strategies. In all cases, three to five levels of increasing assistance were planned. In six studies, a 5 s time delay was inserted before the next level of prompt was administered to allow for independent responding. In three studies, the time delay was not reported, and in three studies, the time delay was variable among levels of prompts. In one case, a 10 s delay was used, and in another, a 30 s delay was used. Prompts consisted primarily of verbal questions or statements, sign, modeling, and partial or full physical guidance or some combination of these. In three studies, a vibration or fan was used to cue responding (Lancioni, O'Reilly, Campodonico, et al. 2000; Lancioni et al., 2007,

Table 3.3 Procedural Parameters for System of Least Prompts

Study	Dependent Variable (Discrete/Chained)	Delay Interval	Prompting Levels	Type of Contingent Reinforcement	BL	Part. Rep.
Berg and Wacker (1989)	Stuff legal envelope (C) Stuff manila envelope (C) Bag 3 items (C)	Not reported	(1) No assistance, (2) partial physical and sign, (3) full physical	Tactile praise (pat on shoulder)	N	1/1 1/1 1/1
Browder et al. (2011)*	Comprehension (D) Engagement (D)	5 s	(1) Verbal, (2) repeat verbal, (3) model, (4) full physical	Verbal praise	N	1/1 1/1
Correa et al. (1984)*	Reach and grasp at midline (D) Reach and grasp at right (D) Reach and grasp at left (D)	10 s	(1) No assistance/toy sound, (2) verbal, (3) model, (4) full physical	Access to toy and social praise (verbal and hugs, pats, or kisses)	Y	3/3 1/1 1/1
Grisham-Brown et al. (2000)*	Remove shirt (C)	5 s	(1) No assistance/coactive sign, (2) partial physical, (3) full physical	Verbal and tactile praise (pat on back), access to activity	Y	1/1
Lancioni et al. (2007)*	Dressing (C)	See levels	(1) No assistance, (2) vibration or verbal after 60 s, (3) full physical after 5–10 s	Verbal praise and tangible (music or vibration)	N	2/2
Lancioni et al. (2008)	Move to activity (D)	See levels	(1) No assistance, (2) fan or verbal after 50 s, (3) full physical after 10–20 s of inactivity or 60–90 s from onset of previous prompt	Tangible (preferred liquid or food)	Y	4/4

Lancioni, O'Reilly, Campodonico, et al. (2000)	Leave laundry area (D) Leave toileting area (D)	See levels	(1) No assistance, (2) vibration after 1 m, (3) touch cue after 3 s, then 5–6 s, then 10 s	Access to toy and tactile praise (hug)	Y	1/1 1/1
Mims et al. (2009)*	Comprehension of books: *Dirty Bertie* (D) *Alexander and the Terrible, Horrible, No Good, Very Bad Day* (D) *Miss you every day* (D)	5 s	(1) No assistance/ comprehension question, (2) verbal, (3) model, (4) full physical	Verbal praise	N	1/1 1/1 1/1
Sisson and Dixon (1986)	Napkin use (D) Utensil use (D)	30 s	(1) No assistance, (2) verbal, (3) full physical	Tangible (access to food) and verbal praise	N	1/1 1/1
Strawbridge et al. (1989)*	Taking steps with walker (D)	5 s	(1) No assistance, (2) touch cue, (3) full physical	None	NA	1/1
Taras et al. (1993)*	Folding (C) Call 911 (C) Spreading with a knife (C) Making coin purse (C) Making belt (C) Making bolo tie (C)	Not reported	(1) No assistance, (2) verbal from instructor or peer, (3) least amount of physical guidance necessary	Verbal praise and tangible (tactile sticker)	N	3/3 3/3 3/3 1/1 1/1 1/1
Taylor (1987)	Prepare cereal (C) Prepare chocolate milk (C) Prepare juice (C)	5 s	(1) No assistance; (2) verbal/ tactile sign, point out salient features on recipe cards; (3) model; (4) full physical	Verbal praise	Y	1/1 1/1 1/1

Continued

Table 3.3 Procedural Parameters for System of Least Prompts—cont'd

Study	Dependent Variable (Discrete/Chained)	Delay Interval	Prompting Levels	Type of Contingent Reinforcement	BL	Part. Rep.
Taylor et al. (1993)	Ordering pizza (C) Renting a movie (C)	5 s	(1) No assistance, (2) verbal/question, (3) verbal assist/clues or reminder, (4) direct verbal instructions, (5) physical assist	Verbal praise	N	2/2 2/2
Trask-Tyler et al. (1994)*	Prepare pizza (C) Prepare coffee (C) Prepare cheesecake (C) Prepare French fries (C)	Not reported	(1) No assistance, (2) verbal, (3) verbal + physical guidance, (4) verbal + hand over hand	None	NA	2/2 3/3 3/3 1/1

BL, contingent reinforcement provided in baseline; Part. Rep., participant replications reported as number of successful attempts out of number of planned attempts; Y, yes; N, no.
*Indicates study meets IES standards with or without reservations.

Table 3.4 Procedural Parameters for Most-to-least Prompts

Study	Dependent Variable (Discrete/Chained)	Number of Prompt Levels and Type of Prompts	Change Criteria	Error Response	Type of Contingent Reinforcement	BL	Part. Rep.
Grisham-Brown et al. (2000)*	Sign "more" (D) Indicate "I want up" (D)	(1) Hand over hand, (2) physical assist at wrist, (3) physical assist at forearm, (4) physical assist at elbow, (5) no assistance (5–10 s)	1–2 Sessions of correct responding	Provide prompt at previous level (# of trials not specified)	Verbal and tactile praise (pat on back), access to activity	Y	1/1 1/1
Gee et al. (1995)*	Grasp/reach (D)	(1) Physical assist to grasp and reach, (2) physical assist to grasp only	80% Accuracy	5 s Delay followed by full physical assistance through all steps of activity	Continuation in routine and verbal praise	N	1/1
Gee et al. (1995)*	Holding objects (D)	(1) Physical assist for 3 s, (2) physical assist for 1 s, (3) no assistance	80% Accuracy	Same as above	Same as above	N	1/1
Gee et al. (1995)*	Maintain arm position (D)	(1) Physical assist to position and hold arm, (2) physical assist to position arm only	80% Accuracy	Same as above	Same as above	N	1/1
Gee et al. (1995)*	Release grasp (D)	(1) Physical assist at wrist for 5 s, (2) physical assist at wrist for 3 s, (3) touch cue at wrist	80% Accuracy	Same as above	Same as above	N	1/1

Continued

Table 3.4 Procedural Parameters for Most-to-least Prompts—cont'd

Study	Dependent Variable (Discrete/Chained)	Number of Prompt Levels and Type of Prompts	Change Criteria	Error Response	Type of Contingent Reinforcement	BL	Part. Rep.
Gee et al. (1995)*	Reach (D)	(1) Physical assist to extend arm 100%, (2) physical assist to extend arm 75%, (3) physical assist to extend arm 50%, (4) no assistance	80% Accuracy	5 s Delay, then full physical assistance through all steps of activity	Continuation in routine and verbal praise	N	1/1
McKelvey et al. (1992)	Put on socks (C) Put on shorts (C) Put on shirt(C)	(1) Hand over hand, (2) thumb and two fingers, (3) thumb and one finger, (4) four fingers, (5) two fingers, (6) one finger, (7) no assistance	10 Consecutive correct trials	Provide prompt at previous level (10 trials)	Tangible (oyster cracker) and social interaction	N	1/1 1/1 1/1
Sisson et al. (1988)*	Put on socks (C) Put on pants (C) Put on shirt (C)	(1) Verbal + hand over hand, (2) verbal + thumb and two fingers, (3) verbal + thumb and one finger, (4) verbal + four fingers, (5) verbal + two fingers, (6) verbal + one finger, (7) verbal + shadowing, (8) verbal only	10 Consecutive correct trials	Provide prompt at previous level (5 trials)	Tangible (candy and crackers) and verbal praise on prompted and unprompted correct responses	N	2/2 2/2 2/2

BL, contingent reinforcement provided in baseline; Part. Rep., participant replications; Y, yes; N, no.
*Indicates study meets IES standards with or without reservations.

Table 3.5 Procedural Parameters for Graduated Guidance

Study	Dependent variable (Discrete/Chained)	Response Interval	Prompting Levels	Fading Procedure	Type of Contingent Reinforcement	BL	Part. Rep.
Lancioni, O'Reilly, Oliva, et al. (2000)	Move to activity (D)	50 s to complete activity and move to next	(1) Verbal, (2) vibration after 10–20 s, (3) touch cues on the back	Touch cues intermittently for 1 m intervals 3–7 s after onset of vibration	Tangible (preferred food) and social interaction	N	2/2
Reese and Snell (1991)	Taking jacket off (C) Putting jacket on (C) Taking coat off (C) Putting coat on (C)	5 s for each step of task analysis	(1) Hand over hand, (2) index finger and thumb, (3) one finger touch, (4) no assistance	Consideration of student responses on previous trials and interpreting pressure cues	Tangible (Cheerios), tactile praise (pats on back), and verbal praise	N	2/2 2/2 2/2 2/2
Wright and Wolery (2012)*	Street crossing at residential two-lane intersection (C) Street crossing at light commercial two lane (C)	Not reported	(1) No assistance; (2) verbal, question or reminder; (3) physical assistance	Prompts only provided in response to errors or pause during travel	Verbal praise	N	3/3 2/2

BL, contingent reinforcement provided in baseline; Part. Rep., participant replications; Y, yes; N, no.
*Indicates study meets IES standards with or without reservations.

Table 3.6 Procedural Parameters for Constant and Progressive Time Delay and Simultaneous Prompts

Study	Dependent Variable (Discrete/Chained)	PP	Delay Interval	Controlling Prompt	# of 0 s Trials	Type of Contingent Reinforcement	BL	Part. Rep.
Collins et al. (1991)[a]	Spoon use (C) Cup use (C)	CTD	3 s	Physical guidance	10	Verbal praise	Y	1/1 1/1
Gee et al. (1991)*	Switch activation to request continuation of routine (D)	PTD	Start at 3 s, increase by 2–3 s	Physical guidance	2–17	Verbal praise and continuation of routine	N	2/2
Grisham-Brown et al. (2000)*	Switch activation (D) Follow direction to "put in" (D)	CTD	5 s	Full physical prompt	3 days (36 trials)	Tactile and verbal praise	Y	1/1 1/1
Grisham-Brown et al. (2000)*	Grasp object (D)	SP	NA	Full physical prompt	Three out of every four trials	Tactile and verbal praise	Y	0/1

[a]Graduated guidance was used implemented on Session 84 for one participant for one step for cup use (response to prompt dependency).
BL, contingent reinforcement provided in baseline; Part. rep., participant replications; Y, yes; N, no.
*Indicates study meets IES standards with or without reservations.

2008). Modeling was used as a type of controlling prompt only within SLP systems. Due to the nature of the sensory disability, across studies, modeling consisted of manual guidance through an activity and then returning the students' hands to the starting position and waiting for a response. Interestingly, none of the researchers used hand-under-hand guidance for modeling.

3.1.2.2 Most-to-Least Prompts

A most-to-least prompting hierarchy was used in four studies with two to eight levels of prompts, always beginning with a full physical assist often referred to as hand over hand. Prompts were faded by changing the position of the prompt (Grisham-Brown et al., 2000), the length of time that prompts were provided (Gee et al., 2005), or the number of fingers used to provide physical assistance (McKelvey et al., 1992; Sisson et al., 1988). To move to a lower prompt level, participants were required to respond correctly (i.e., without resistance) for a set number of trials. In three out of four studies, prompts were provided at the previous level in case of an error. Gee et al. (1995) responded to all errors with full manual guidance. Gee et al. also physically guided participants through the entire routine for 10–15 days before implementing the system of prompts.

3.1.2.3 Graduated Guidance

GG was used in just three studies: two using an increasing hierarchy of prompts and one using decreasing levels of prompts. Although assistance was increased, Wright and Wolery (2012) did not structure their system of prompts in a rigid format. They did not specify the length of time delay before administering prompts and allowed the instructor to determine the exact level of assistance needed according to the students' responses. Lancioni, O'Reilly, Oliva, et al. (2000) used a more prescribed approach. After an initial time delay, they provided a verbal task direction, and then, if participants did not respond after 10–20 s, a vibratory cue was provided. After onset of the vibratory cue, participants were provided touch cues intermittently as necessary until they reached their destination. This procedure fits the definition for GG because the level of prompting is fluid and can increase or decrease during the activity. To administer GG in a decreasing fashion, first participants were provided an opportunity to complete the task independently, and then, physical assistance was provided and faded over time (Reese & Snell, 1991). Prompts were faded by level of intensity (i.e., using two fingers and then one finger) from the interventionist's interpretation of pressure cues.

3.1.2.4 Time Delay

CTD was used in two studies, PTD in one and SP in one study. For time delay, a 3- to 5-s time delay was inserted after 0 s trials, and a full physical prompt was used as the controlling prompt. The number of 0 s trials provided varied across studies (range, 2–36 trials). SP prompting was used in only one instance to teach grasping and was unsuccessful (Grisham-Brown et al., 2000).

3.2. Participant Characteristics and Study Contexts

3.2.1 Settings

Study characteristics including the settings for instruction can be found in Table 3.7. Research was primarily carried out during daily routines in a school or residential facility in a special setting for students or adults with disabilities ($n=15$), an inclusive classroom ($n=3$), the community ($n=3$), or the home ($n=1$). The specific location for instruction was always appropriate and functional for the skills being taught. For example, dressing skills were taught in participants' bedrooms (Lancioni et al., 2007; Sisson et al., 1988); mobility was taught in the hallway, in the street, or in the transition between activities (Lancioni, O'Reilly, Campodonico, et al., 2000; Lancioni et al., 2008; Strawbridge et al., 1989; Wright & Wolery, 2012); meal preparation was taught in the kitchen (Taylor, 1987; Trask-Tyler et al., 1994); object communication to demonstrate understanding of books was taught during shared story time (Browder et al., 2011; Mims et al., 2009); and communication and motor skills were taught in functional routines or in a play context (Correa et al., 1984; Gee et al., 1991, 1995; Grisham-Brown et al., 2000). All participants received one-to-one instruction except six students who participated in small-group instruction (Reese & Snell, 1991; Taras et al., 1993). Berg and Wacker (1989) and Lancioni, O'Reilly, Oliva, et al. (2000) did not give the exact instructional setting. The independent variable was implemented by school staff in seven studies (31.8%) and by researchers with an unknown relationship to the participants in seven studies (31.8%). In five studies, the interventionist was not described.

3.2.2 Participants

Participant characteristics are provided in Table 3.8. The mean age of participants was 13.6 years old (range: 2 years and 3 months to 43 years). The breakdown of participants by age is as follows: birth to 5 years ($n=8$), 6–12 years ($n=14$), 13–18 years ($n=7$), and 19 or older ($n=13$). Time delay

Table 3.7 Study Characteristics

Study	Journal	No.	Location	DV Construct	Prompting System	Design	Mn	Gen	SV
Berg and Wacker (1989)	J Appl Behav Anal	1	Not reported	Stuffing envelopes and bagging	SLP	MB-B	Y	Y	N
Browder et al. (2011)*	Educ Train Dev Disabil	1	Self-contained classroom	Comprehension and engagement	SLP	MP-P	Y	N	Y
Collins et al. (1991)	J Dev Phys Disabil	1	Self-contained classroom	Spoon and cup use	CTD GG[a]	MP-B	Y	Y	N
Correa et al. (1984)*	J Appl Behav Anal	3	Self-contained classroom	Reach and grasp	SLP[b]	MB-P MB-B	N	Y	N
Gee et al. (1991)*	J Assoc Pers Sev Handicaps	2	Inclusive classroom	Requesting continuation of routine	PTD	MP-P	N	Y	N
Gee et al. (1995)*	Behav Modif	2	Inclusive classroom and community	Functional motor skills	MLP	MP-B	N	N	N
Grisham-Brown et al. (2000)*	J Behav Educ	3	Inclusive classroom	Switch activation and functional motor skills	SLP MLP CTD SP	MP-P	N	N	N
Lancioni et al. (2007)*	J Vis Impair Blind	2	Self-contained residential setting	Dressing	SLP	MB-P	N	N	N
Lancioni et al. (2008)	Res Dev Disabil	4	Self-contained residential setting	Independent travel to activities	SLP	MB-P	N	N	N

Continued

Table 3.7 Study Characteristics—cont'd

Study	Journal	No.	Location	DV Construct	Prompting System	Design	Mn	Gen	SV
Lancioni, O'Reilly, Campodonico, et al. (2000)	Scand J Behav Ther	1	Self-contained school	Leaving laundry or toilet area	SLP	MP-B	Y	N	N
Lancioni, O'Reilly, Oliva, et al. (2000)	Scand J Behav Ther	2	Self-contained residential setting	Ambulation	GG	MB-P	N	N	N
McKelvey et al. (1992)	Teach Excep Child	1	Bedroom (self-contained residential setting)	Dressing	MLP[b]	MP-B	Y	N	N
Mims et al. (2009)*	Educ Train Dev Disabil	1	Self-contained classroom	Reading comprehension	SLP	MP-B	Y	Y	Y
Reese and Snell (1991)	J Vis Impair Blind	2	Self-contained classroom	Putting on and taking off coat/ jacket	GG	MP-B	Y	N	N
Sisson and Dixon (1986)	J Vis Impair Blind	1	Self-contained classroom	Napkin and utensil use	SLP	MB-B	N	N	N
Sisson et al. (1988)*	Res Dev Disabl	2	Bedroom (self-contained residential setting)	Dressing skills	MLP[b]	MB-B	Y	Y	N
Strawbridge et al. (1989)*	Behav Ther	1	School hallway	Taking steps	SLP	Withdrawal	Y	N	N

Taras et al. (1993)*	Behav Modif	4	Self-contained classroom	Daily living skills and leather making	SLP	MB-B	Y	N	Y
Taylor (1987)	J Vis Impair Blind	1	School kitchen	Meal preparation	SLP	MP-B	N	N	N
Taylor et al. (1993)	J Behav Educ	2	School office, community	Leisure skills	SLP	MB-P	Y	N	N
Trask-Tyler et al. (1994)*	J Behav Educ	3	Kitchen (self-contained residential setting)	Food preparation	SLP	MP-B	Y	Y	N
Wright and Wolery (2012)*	J Spec Educ	3	Community	Street crossing	GG	MP-P	Y	Y	N

[a]Graduated guidance was used implemented on Session 84 for one participant for one step for cup use (response to prompt dependency).
[b]Prompting system is different than what the authors call it.
No., the number of participants with visual impairment that meets inclusion criteria (journals are listed by their NLM abbreviations); DV, dependent variable; SLP, system of least prompts; MLP, most-to-least prompting; GG, graduated guidance; CTD, constant time delay; SP, simultaneous prompting; MP-P (B), multiple probe across participants (behaviors); MB-P (B), multiple baseline across participants (behaviors); Gen, generalization; Mn, maintenance; SV, social validity; Y, yes; N, no.
*Indicates study meets IES standards with or without reservations.

Table 3.8 Participant Characteristics

Study	Age	Sex	Diagnoses	Functional Description
Berg and Wacker (1989)	19	F	NR	DB, NLP, verbal, ambulatory, moderate ID
Browder et al. (2011)*	9	F	NR	RV, NV, noncompliant behavior, severe ID
Collins et al. (1991)	3.8	F	CVI, hypotonic CP	No RV, NV, NA, severe ID
Correa et al. (1984)*	2.4	M	CVI, CP	LP, NV, SD, NA, severe ID
	4.3	M	ONA, CVI	NLP, congenital, NV, NA, SD, stereotypical behaviors, severe ID
	2.3	M	Hydrocephalus	LP, congenital, NV, SD, NA, severe ID
Gee et al. (1991)*	7		NR	NLP, NV, SD, NA, profound ID
	10		Gastrointestinal degenerative disorder	LP, NV, SD, NA, profound ID
Gee et al. (1995)*	9	M	Quadriplegia	DB, no RV, NV, SD, NA, severe ID
	11	M	Quadriplegia	DB, LP, severe field losses, NV, SD, NA, severe ID
Grisham-Brown et al. (2000)*	4.4	M	NR	DB, NV, ambulatory, severe MD, SIB, severe ID
	3.10	M	CP, strabismus	DB, no RV, NV, SD, NA, TD, severe ID
	2.4		NR	DB, NV, SD, ambulatory, stereotypical behavior, severe ID
Lancioni et al. (2007)*	27	M	Encephalopathy	No RV, ambulatory, profound ID
	29	M	Encephalopathy	DB, no RV, NA, profound ID

Lancioni et al. (2008)	43	M	NR	DB, no RV, NV, NA, profound ID
	27	M	NR	No RV, NV, NA, profound ID
	20	M	NR	No RV, NV, NA, profound ID
	22	M	NR	DB, RV, NV, NA, profound ID
Lancioni, O'Reilly, Campodonico, et al. (2000)	14.5	M	NR	DB, LP, ambulatory, stereotypical behavior, profound ID
Lancioni, O'Reilly, Oliva, et al. (2000)	20	M	NR	No RV, ambulatory, stereotypical behavior, EBD, profound ID
	21	M	NR	No RV, ambulatory, stereotypical behavior, EBD, profound ID
McKelvey et al. (1992)	15	F	CVI	DB, LP, profound bilateral HL, SD, ambulatory, noncompliant behavior, stereotypical behavior, severe ID
Mims et al. (2009)*	6	M	CP, CVI, bronchopulmonary dysplasia	NA, severe ID
Reese and Snell (1991)	6	F	NR	DB, no RV, NV, NA, severe ID
	6	F	Leber syndrome	No RV, verbal, SD, NA, moderate ID
Sisson and Dixon (1986)	10	F	NR	No RV, congenital, NV, EBD, severe ID
Sisson et al. (1988)*	9	M	De Morsier syndrome, panhypopituitarism, diabetes insipidus	LP, ambulatory, SD, noncompliant behavior, stereotypical behavior, moderate ID
	5	M	Prematurity, hydrocephalus, CP	LP, SD, severe MD, noncompliant behavior, severe ID

Continued

Table 3.8 Participant Characteristics—cont'd

Study	Age	Sex	Diagnoses	Functional Description
Strawbridge et al. (1989)*	9	M	ONA	DB, LP, severe bilateral HL, NV, NA, stereotypical behavior, profound ID
Taras et al. (1993)*	8	M	NR	LP, mild ID
	7	M	NR	RV, verbal, no ID
	9	F	NR	LP, verbal, mild ID
	19	M	NR	LP, mild ID
Taylor (1987)	21	F	NR	DB, no RV, NV, ambulatory, profound ID
Taylor et al. (1993)	16	M	Cataracts, glaucoma, RLF, CP	No RV, ambulatory, EBD, noncompliant behavior, mild ID
	16	M	RLF, microphthalmia, scoliosis	DB, No RV, ambulatory, EBD, mild ID
Trask-Tyler et al. (1994)*	21	F	NR	No RV, verbal, echolalic, mild ID
	20	M	NR	No RV, verbal, mild ID
	17	M	Degenerative neurological disease	No RV, verbal, SD, TD, mild ID
Wright and Wolery (2012)*	16.4	F	Brain tumor	LP, adventitious, verbal, ambulatory, no ID
	14.10	M	ROP	DB, NLP, mild HL, congenital, verbal, ambulatory, no ID
	20	F	ONH	NLP, congenital, verbal, ambulatory, no ID

Note: Age is reported in years and months. EBD, emotional or behavioral disorder; NV, nonverbal; ID, intellectual disability; RV, residual vision (minimal); LP, light perception; SD, seizure disorder; NLP, no light perception; NA, nonambulatory; TD, technology-dependent; RLF, retrolental fibroplasia; ONH, optic nerve hypoplasia; ONA, optic nerve atrophy; DB, deafblind; CP, cerebral palsy; HL, hearing loss.

*Indicates study meets IES standards with or without reservations.

strategies were used with the youngest students to teach spoon and cup use (Collins et al., 1991) and discrete motor or communication skills (Grisham-Brown et al., 2000). Other preschool children were taught discrete or chained motor skills using SLP or MLP (Correa et al., 1984; Grisham-Brown et al., 2000). Elementary and middle school students were taught self-dressing with MLP or GG (Reese & Snell, 1991; Sisson et al., 1988) and other self-care or daily living skills using SLP (Sisson & Dixon, 1986; Taras et al., 1993); communication skills with SLP or PTD (Browder et al., 2011; Gee et al., 1991; Mims et al., 2009); or discrete motor skills with MLP or SLP (Gee et al., 1995; Strawbridge et al., 1989). All but four students (Taras et al., 1993) under the age of 12 had moderate to profound intellectual disabilities. High school-age participants under 22 years of age were taught self-dressing (McKelvey et al., 1992), mobility (Lancioni, O'Reilly, Campodonico, et al., 2000; Lancioni, O'Reilly, Oliva, et al., 2000; Wright & Wolery, 2012), leisure/vocational skills (Berg & Wacker, 1989; Taras et al., 1993; Taylor et al., 1993), or meal preparation (Taylor, 1987; Trask-Tyler et al., 1994). Older adults were taught dressing (Lancioni et al., 2007) and mobility (Lancioni et al., 2008) using SLP.

Fewer than half of the study sample reported visual diagnoses (Collins et al., 1991; Correa et al., 1984; McKelvey et al., 1992; Mims et al., 2009; Reese & Snell, 1991; Sisson et al., 1988; Strawbridge et al., 1989; Taylor et al., 1993; Wright & Wolery, 2012). Of the visual diagnoses represented, the most common were retinal diseases ($n=5$), cortical visual impairment (CVI; $n=5$), and optic nerve disorders ($n=4$). Additionally, this sample included students diagnosed with glaucoma ($n=1$), microphthalmia ($n=1$), cataracts ($n=2$), and strabismus ($n=1$). For 29 participants, no visual diagnoses were provided, and for 21 of those participants, no medical diagnoses were reported either. The most common medical diagnoses were cerebral palsy ($n=6$), hydrocephalus or encephalopathy ($n=4$), and quadriplegia ($n=2$). Sixteen participants were deafblind, but only two participants' hearing loss was described (McKelvey et al., 1992; Strawbridge et al., 1989). In only two studies was participants' onset of vision loss described (Sisson & Dixon, 1986; Wright & Wolery, 2012). Sufficient information was provided to determine that 4 participants did and 35 did not have residual vision; 16 participants had light perception and 5 did not. Insufficient information was provided for 22 participants to know whether they had light perception.

All but four participants (Taras et al., 1993; Wright & Wolery, 2012) had intellectual disability (ID). Eight participants had mild ID (Taras et al., 1993;

Taylor et al., 1993; Trask-Tyler et al., 1994), and only one student had moderate ID (Berg & Wacker, 1989). Sixteen participants were identified as nonverbal, 10 as verbal, and the verbal abilities of 17 were not described. In studies that reported sufficient information to determine whether students were ambulatory, 15 were and 18 were not. Additionally, 14 participants had seizures disorders, 5 had an emotional or behavioral disorder, and 8 had stereotypical behaviors. Males ($n=31$) made up 72% of the sample.

3.3. Determination of Evidence-Based Practice

A summary of the extent to which each study met the Horner et al. (2005) quality indicators for single-subject research is provided in Table 3.1. Four studies met standards without reservations (Browder et al., 2011; Gee et al., 1991; Mims et al., 2009; Wright & Wolery, 2012). Eight more studies met standards with reservations (Correa et al., 1984; Gee et al., 1995; Grisham-Brown et al., 2000; Lancioni et al., 2007; Sisson et al., 1988; Strawbridge et al., 1989; Taras et al., 1993; Trask-Tyler et al., 1994). These 12 studies originated from eight independent research groups as defined by Ledford et al. (2012). Across the studies that met standards, there were 57 replications of effect among 27 participants who met the inclusion criteria for this chapter. Therefore, response prompting appears to have enough support at present to qualify as an EBP to teach skill acquisition to students with little to no functional vision. Presently, we have the most research evidence to support the use of SLP. Across 17 participants and 21 behaviors, there were 35 successful replications of the effect of SLP on skill acquisition. There were only 16 replications across six participants using MLP, five replications across three participants using GG, two replications within one participant for CTD, two replications within one participant for PTD, and no replications using SP. Of the studies meeting standards, the maintenance of skill acquisition was only demonstrated for SLP, MLP, and GG in seven studies. Likewise, the generalization of skill acquisition was demonstrated for SLP, MLP, PTD, and GG; however, these data were collected in only 50% of studies meeting standards.

4. DISCUSSION

This chapter was conducted to synthesize and describe the literature on response prompting to teach new skills to learners with blindness. Specifically, the questions that guided this endeavor were the following: (a) What types of prompting strategies are being used to teach what skills under what conditions? (b) What are the characteristics of participants with little to no functional vision (i.e., blindness) for whom response prompting has been used to teach skill acquisition? (c) Is response prompting an evidence-based practice for teaching functional skills to students with blindness?

4.1. Prompting Strategies, Target Skills, and Contexts

Twenty-two studies that used prescriptive response-prompting procedures were evaluated; those studies employed most-to-least prompts, GG, system of least prompts, constant time delay, progressive time delay, and simultaneous prompts to teach functional skills. Although both chained and discrete skills were taught using time delay and hierarchies, time delay was used to teach chained skills in only one study, which did not meet quality indicator standards. Prompting procedures in general were used for a variety of skills taught within a variety of contexts in schools and community settings. Most often targeted skills were embedded within natural routines with multiple opportunities to respond within and across sessions. Most of the research supports the use of prompting for this group of learners in one-to-one instructional settings with an experienced teacher or familiar adult. Few studies reported teachers' prior experience or credentials to work with students who are blind (e.g., Wright & Wolery, 2012).

Procedural parameters for interventions mostly matched the descriptions presented by Wolery and Gast (1984) except in the case of GG, which in some cases more closely resembled MLP (McKelvey et al., 1992; Sisson et al., 1988) or SLP (Correa et al., 1984). Despite the procedure (i.e., GG and MLP), prompting was often physical, in the form of hand-over-hand guidance. System of least prompts was used most often, and modeling was often used as an intermediary level within SLP systems. Unlike studies with sighted participants, modeling always involved manual guidance. Additionally, in several studies, manual guidance was provided to students to model task completion prior to starting intervention (Gee et al., 1995; Sisson & Dixon, 1986; Taras et al., 1993).

For time delay, 0 s trials were generally administered until students achieved 80% correct responding; thereafter, a 5 s time delay was inserted with a full physical prompt as the controlling prompt. This criterion level is lower than that typically set for students with vision (i.e., 100% at 0 s). The number of 0 s trials should correlate with task difficulty and functioning level of the learner (Wolery & Gast, 1984). Participants in this sample generally required more 0 s delay trials than sighted participants with severe disabilities in studies reviewed by Dogoe and Banda (2009).

4.2. Participant Characteristics

Based on the functional descriptions of participants, the findings from this chapter can be generalized to students with blindness who require intensive supports in school. Participants presented with multiple disabilities primarily, including intellectual and physical disabilities in most cases. A broad range of ages, visual disabilities, and causes of disability were represented. Time delay strategies in particular were used exclusively with young children under the age of 10. The three most common visual diagnoses (i.e., CVI, ROP, and ONH) were represented in the sample. Learners with deafblindness were also common. Response-prompting procedures were effective for old and young learners with challenges in all developmental domains, including physical, emotional, adaptive behavior, communication, social, and cognitive.

4.3. Progress Toward Identifying an Evidence-based Practice

Unfortunately, the process for identifying a practice as an EBP is challenging. The Horner et al. (2005) criteria are frequently used to evaluate the single-subject literature, but a replicable procedure for applying the criteria has not emerged, making comparison across reviews difficult (Cook & Cook, 2013). Additionally, different search methods and inclusion criteria add to the difficulty of comparison. Nevertheless, it may be beneficial to compare the rate with which other researchers have identified studies as meeting quality indicators to get a sense of how liberal or conservative the judgments are for this chapter. Although similar search criteria were used, Browder et al. (2009) described 73.3% of the 30 studies as high quality to identify an EBP for literacy instruction for students with significant cognitive disabilities. In our study, we described 54.5% of the 22 studies as being high quality. Similarly, Ledford et al. (2012) identified 47 studies for a review of time delay procedures to teach discrete skills during small-group instruction and used findings from all 47 for identifying this EBP. Within the area of visual impairments, reviews that reference the Horner et al. criteria do not systematically apply the criteria to evaluate the quality of the literature. Taken together, this evidence suggests that our assessment of the quality of the studies in this chapter was conservative.

Twelve of the 22 studies met high quality standards according to the Horner et al. (2005) quality indicators for single-subject research. Eight different research groups completed these studies. Across the 12 studies, 27 participants with little to no functional vision were included. Skills taught to those 27 participants included comprehension and engagement during story time, functional motor skills, switch activation for communication, dressing skills, meal preparation, and street crossing. A sufficient number of intersubject replications were documented to qualify response prompting as an evidence-based practice to teach skill acquisition to learners who are blind.

4.4. Implications for Practice

Response-prompting procedures are being used in classrooms serving students with blindness and multiple disabilities, including students who are nonambulatory and nonverbal, and should continue to be used, and their use should be expanded. Target skills may be embedded in natural routines, and the continuation of the activity may be a sufficient reinforcer to increase the likelihood the behavior will occur in the future. Although we found more evidence for hierarchical prompting systems than for time delay procedures, we need more research on time delay. Time delay may be more efficient than hierarchical prompting systems to teach students with other disabilities discrete skills. There was also more evidence for increasing levels of prompting (i.e., SLP) over decreasing levels (i.e., most to least); however, modeling in the form of manual guidance may be needed to preteach the skill sequence prior to instruction with an SLP system. Target skills should be selected that are likely to be achieved in less than a month, given daily opportunities to respond. These may be very specific functional motor movements or long chained skills, depending on the student.

4.5. Directions for Future Research

Much more research is needed to equip professionals with the information required to select and design effective and efficient prompting systems to use with their students. More rigorous research in this domain is being conducted. Specifically, as shown in Table 3.1, the three most recently published studies (Browder et al., 2011; Mims et al., 2009; Wright & Wolery, 2012) met all of the Horner et al. (2005) quality indicators. Future research should continue in this direction to improve this body of research by collecting fidelity data as well as generalization, maintenance, and social validity data. In addition to improving the quality of research, systematic replications by at least three independent research groups should be carried out to further define for whom and under what conditions response prompting is effective in a variety of contexts.

Based on other systematic reviews of response-prompting procedures, there are two specific areas to expand the research on prompting for students with blindness. First, because response prompting has been identified as an EBP in the context of small-group instruction for students with disabilities in general, this may be a logical direction to focus intervention research efforts. Small-group instructional strategies would likely require thoughtful adaptation to be meaningful for students without sight. Second, time delay, although seldom used in the studies we reviewed, has a large evidence base to support its use to teach literacy to students with disabilities who are sighted. Therefore, time delay may be effective in teaching communication, language, braille contractions, and other literacy and academic skills to students without vision. Intervention research in this area is also needed.

4.6. Limitations

There are a few potential limitations to consider when interpreting the results of this study. First, the results of this chapter do not isolate the effects of prompting without the delivery of various types of reinforcement. In fact, reinforcement was not always available in baseline. Therefore, the changes in the dependent measures for those studies may be due to reinforcement procedures. Second, although our findings show that sufficient evidence has been published to grant evidence-based practice status to response prompting for learners with blindness, readers are cautioned to think critically. The participants in this set of studies span a wide age range and ability level and were taught a wide range of tasks of varying levels of difficulty. Additionally, no two prompting procedures were exactly alike. An evidence-based practice may be defined narrowly or broadly in terms of the specific group and specific set of conditions to which it applies. In this case, the group was defined as learners with blindness (most of whom had multiple disabilities) and the target behavior as skill acquisition. Still, much variability exists within these parameters.

4.7. Conclusion

Wolery and Gast (2000) highlighted the importance of reading the literature from other fields and disciplines. By reviewing literature from special education, early childhood, and applied behavior analysis, much has been learned about strategies being used with students who are blind, and areas of need have been identified. Specifically, sufficient research has been published to support the use of response-prompting procedures to teach skill acquisition to learners with blindness. Presently, response prompting is a tool at our disposal to teach students with the most significant needs; however, there is little research to guide instructional planning. Systematic replications will provide information about what prompting procedures are most efficient for specific contexts.

REFERENCES

Akmanoglu-Uludag, N., & Batu, S. (2005). Teaching naming relatives to individuals with autism using simultaneous prompting. *Education and Training in Developmental Disabilities*, *40*(4), 401–410.

Ayres, K., & Gast, D. L. (2010). Dependent measures and measurement procedures. In D. Gast (Ed.), *Single subject research methodology in behavioral sciences* (pp. 129–165). New York, NY: Routledge.

Bennett, D. L., Gast, D. L., & Schuster, J. W. (1986). Time delay and system of least prompts: A comparison in teaching manual sign production. *Education and Training of the Mentally Retarded*, *21*, 117–129.

Berg, W. K., & Wacker, D. P. (1989). Evaluation of tactile prompts with a student who is deaf, blind, and mentally retarded. *Journal of Applied Behavior Analysis*, *22*, 93–99.

Bigelow, A. (1995). The effect of blindness on the early development of the self. In P. Rochat (Ed.), *The self in infancy: Theory and research: Vol. 112* (pp. 327–347). Amsterdam: Elsevier.

Bozkurt, F., & Gursel, O. (2005). Effectiveness of constant time delay on teaching snack and drink preparation skills to children with mental retardation. *Education and Training in Developmental Disabilities*, *40*(4), 390–400.

Brambring, M. (2007). Divergent development of manual skills in children who are blind or sighted. *Journal of Visual Impairment & Blindness*, *101*, 212–225.

Browder, D., Ahlgrim-Delzell, L., Spooner, F., Mims, P. J., & Baker, J. N. (2009). Using time delay to teach literacy to students with severe developmental disabilities. *Exceptional Children*, *75*, 343–364.

Browder, D. M., Lee, A., & Mims, P. (2011). Using shared stories and individual response modes to promote comprehension and engagement in literacy for students with multiple, severe disabilities. *Education and Training in Autism and Developmental Disabilities*, *46*, 339–351.

Brown, L., Bellamy, T., Perlmutter, L., Sackowitz, P., & Sontag, E. (1972). The development of quality, quantity, and durability in the work performance of retarded students in a public school prevocational workshop. *Training School Bulletin*, *68*, 58–69.

Bruce, S., & Muhammad, Z. (2009). The development of object permanence in children with intellectual disability, physical disability, autism, and blindness. *Journal of Disability, Development and Education*, *56*, 229–246.

Çiftçi, H. D., & Temel, Z. F. (2010). A comparison of individual and small-group instruction with simultaneous prompting for teaching the concept of color to children with a mental disability. *Social Behavior and Personality*, *38*(4), 479–493.

Cohen, E. T., Heller, K. W., Alberto, P., & Fredrick, L. D. (2008). Using a three-step decoding strategy with constant time delay to teach word reading to students with mild and moderate mental retardation. *Focus on Autism and Other Developmental Disabilities*, *23*(2), 67–78.

Coleman-Martin, M. B., & Heller, K. W. (2004). Using a modified constant prompt-delay procedure to teach spelling to students with physical disabilities. *Journal of Applied Behavior Analysis*, *37*, 469–480.

Collins, B., Evans, A., Creech-Galloway, C., Karl, J., & Miller, A. (2007). Comparison of the acquisition and maintenance of teaching functional and core content sight words in special and general education settings. *Focus on Autism and Other Developmental Disabilities, 22*, 220–233.

Collins, B. C., Gast, D. L., Wolery, M., Holcombe, A., & Leatherby, J. G. (1991). Using constant time delay to teach self-feeding to young students with severe/profound handicaps: Evidence of limited effectiveness. *Journal of Developmental and Physical Disabilities, 3*, 157–179.

Cook, B. G., & Cook, S. C. (2013). Unraveling evidence-based practices in special education. *Journal of Special Education, 47*(2), 71–82. http://dx.doi.org/10.1177/0022466911420877.

Cook, B. G., & Odom, S. L. (2013). Evidence-based practices and implementation science in special education. *Exceptional Children, 79*, 135–144.

Correa, V., Poulson, C., & Salzberg, C. (1984). Training and generalization of reach-grasp behavior in blind retarded young children. *Journal of Applied Behavior Analysis, 17*, 56–69.

Cosbey, J. E., & Johnston, S. (2006). Using a single-switch voice output communication aid to increase social access for children with severe disabilities in inclusive classrooms. *Research and Practice for Persons with Severe Disabilities, 31*, 144–156.

DeMario, N. C., & Crowley, E. P. (1994). Using applied behavior analysis procedures to change the behavior of students with visual disabilities: A research review. *Journal of Visual Impairment & Blindness, 88*, 532–543.

Demchak, M. (1990). Response prompting and fading methods: A review. *American Journal on Mental Retardation, 94*, 603–615.

Dipipi, C. M., Jitendra, A. K., & Miller, J. A. (2001). Reducing Repetitive Speech: Effects of Strategy Instruction. *Preventing School Failure, 45*(4), 177–181.

DiPipi-Hoy, C., & Jitendra, A. (2004). A parent-delivered intervention to teach purchasing skills to young adults with disabilities. *The Journal of Special Education, 38*(3), 144–157.

Dogan, O. S., & Tekin-Iftar, E. (2002). The effects of simultaneous prompting on teaching receptively identifying occupations from picture cards. *Research in Developmental Disabilities, 23*, 237–252.

Dogoe, M., & Banda, D. R. (2009). Review of recent research using constant time delay to teach chained tasks to persons with developmental disabilities. *Education and Training in Developmental Disabilities, 44*, 177–186.

Dogoe, M. S., Banda, D. R., Lock, R. H., & Feinstein, R. (2011). Teaching generalized reading of product warning labels to young adults with autism using the constant time delay procedure. *Education and Training in Autism and Developmental Disabilities, 46*(2), 204–213.

Doyle, P. M., Wolery, M., Halle, J., & Gast, D. L. (1988). System of least prompts: A review of procedural parameters. *Journal of the Association for Persons with Severe Handicaps, 13*, 28–40.

Dyches, T. T., Davis, A., Lucido, B. R., & Young, J. R. (2002). Generalization of skills using pictographic and voice output communication devices. *Augmentative and Alternative Communication, 18*(2), 124–131.

Erin, J. N. (2003). *Educating students with visual impairments (ERIC EC Digest #E653).* The ERIC Clearinghouse on Disabilities and Gifted Education, Retrieved from http://www.hoagiesgifted.org/eric/e653.html.

Ferrell, K. A. (2006). Evidence-based practices for students with visual disabilities. *Communication Disorders Quarterly, 28*, 42–48.

Fiscus, R. S., Morse, T. E., Schuster, J. W., & Collins, B. C. (2002). Teaching elementary students with cognitive disabilities food preparation skills while embedding instructive feedback in the prompt and consequent event. *Education and Training in Mental Retardation and Developmental Disabilities, 37*(1), 55–69.

Fraiberg, S. (1977). *Insights from the blind: Comparative studies on blind and sighted infants.* New York, NY: Basic Books.

Gee, K., Graham, N., Goetz, L., Oshima, G., & Yoshioka, K. (1991). Teaching students to request the continuation of routine activities by using time delay and decreasing physical assistance in the context of chain interruption. *Journal of the Association for Persons with Severe Handicaps*, *16*, 154–167.

Gee, K., Graham, N., Sailor, W., & Goetz, L. (1995). Use of integrated, general education, and community settings as primary contexts for skill instruction for students with severe, multiple disabilities. *Behavior Modification*, *19*, 33–58.

Grisham-Brown, J., Schuster, J. W., Hemmeter, M. L., & Collins, B. C. (2000). Using an embedding strategy to teach preschoolers with significant disabilities. *Journal of Behavioral Education*, *10*, 139–162.

Hamre-Nietupski, S., Nietupski, J., Sandvig, R., & Ayres, B. (1984). Leisure skills instruction in a community residential setting with young adults who are deaf/blind severely handicapped. *The Journal of the Association for Persons with Severe Handicaps*, *9*, 49–54.

Handen, B. L., & Zane, T. (1987). Delayed prompting: A review of procedural variations and results. *Research in Developmental Disabilities*, *8*, 307–330.

Hatton, D. D., Bailey, D. B., Burchinal, M. R., & Ferrell, K. A. (1997). Developmental growth curves of preschool children with vision impairments. *Child Development*, *68*, 788–806.

Hatton, D. D., Ivy, S. E., & Boyer, C. (2013). Severe visual impairments in infants and toddlers in the United States. *Journal of Visual Impairment & Blindness*, *107*(5), 325–336.

Horner, R. H., Carr, E. G., Halle, J., McGee, G., Odom, S., & Wolery, M. (2005). The use of single-subject research to identify evidence-based practice in special education. *Exceptional Children*, *71*, 165–179.

Kratochwill, T. R., Hitchcock, J., Horner, R. H., Levin, J. R., Odom, S. L., Rindskopf, D. M., et al. (2010). *Single-case designs technical documentation,* Retrieved from What Works Clearinghouse Website http://ies.ed.gov/ncee/wwc/pdf/wwc_scd.pdf.

Lancioni, G. E., O'Reilly, M. F., Campodonico, F., & Serenelli, S. (2000). Use of anticipatory cues to reduce dependence on physical prompts by an adolescent with multiple disabilities. *Scandinavian Journal of Behaviour Therapy*, *29*, 43–47.

Lancioni, G. E., O'Reilly, M. F., Oliva, D., Bianchi, M., & Pirani, P. (2000). Promoting functional ambulation with people with blindness and multiple disabilities. *Scandinavian Journal of Behaviour Therapy*, *29*, 148–151.

Lancioni, G. E., O'Reilly, M. F., Singh, N. N., Sigafoos, J., Oliva, D., Campodonico, F., et al. (2007). Helping three persons with multiple disabilities acquire independent dressing through assistive technology. *Journal of Visual Impairment & Blindness*, *101*, 768–773.

Lancioni, G. E., Singh, N. N., O'Reilly, M. F., Sigafoos, J., Oliva, D., Campodonico, F., et al. (2008). Assisting persons with multiple disabilities to move through simple occupational activities with automatic prompting. *Research in Developmental Disabilities*, *29*, 439–446.

Ledford, J. R., Lane, J. D., Elam, K. L., & Wolery, M. (2012). Using response-prompting procedures during small-group direct instruction: Outcomes and procedural variations. *American Journal on Intellectual and Developmental Disabilities*, *117*, 413–434.

Libby, M. E., Weiss, J. S., Bancroft, S., & Ahearn, W. H. (2008). A comparison of most-to-least and least-to-most prompting on the acquisition of solitary play skills. *Behavior Analysis in Practice*, *1*, 37–43.

Luiselli, J. K. (1993). Training self-feeding skills in children who are deaf and blind. *Behavior Modification*, *17*, 457–473.

McDonnel, J., & Ferguson, B. (1989). A comparison of time delay and decreasing prompt hierarchy strategies in teaching banking skills to students with moderate handicaps. *Journal of Applied Behavior Analysis*, *22*, 85–91.

McKelvey, J. L., Sisson, L. A., Van Hasselt, V. B., & Hersen, M. (1992). An approach to teaching self-dressing to a child with dual sensory impairment. *Teaching Exceptional Children, 25*, 12–15.

Mims, P. J., Browder, D. M., Baker, J. N., Lee, A., & Spooner, F. (2009). Increasing comprehension of students with significant intellectual disabilities and visual impairments during shared stories. *Education and Training in Developmental Disabilities, 44*, 409–420.

Morse, T. E., & Schuster, J. W. (2004). Simultaneous prompting: A review of the literature. *Education and Training in Developmental Disabilities, 39*, 153–168.

National Professional Center on Autism Spectrum Disorders. (2010). *Evidence-based practice: Prompting*, Retrieved from http://autismpdc.fpg.unc.edu/content/prompting.

Neitzel, J., & Wolery, M. (2009). *Steps for implementation: Graduated guidance*. Chapel Hill, NC: The National Professional Development Center on Autism Spectrum Disorders, FPG Child Development Institute, The University of North Carolina.

Odom, S., Dunst, C., Horner, R., & Bradley, R. (2012). Single-case research in OSEP projects: Emerging standards for design, interpretation, synthesis, and meta-analysis. In *Paper presented at the OSEP Project Directors' Conference, Washington, DC.*

Parker, A. T., Davidson, R., & Banda, D. R. (2007). Emerging evidence in single-subject research in the field of deaf-blindness. *Journal of Visual Impairment & Blindness, 101*, 690–700.

Parker, A. T., & Pogrund, R. L. (2009). A review of research on the literacy of students with visual impairments and additional disabilities. *Journal of Visual Impairment & Blindness, 103*, 635–648.

Preisler, G. M. (1993). A descriptive study of blind children in nurseries with sighted children. *Child: Care, Health and Development, 19*, 295–315.

Realon, R. E., Favell, J. E., & Dayvault, K. A. (1988). Evaluating the use of adaptive leisure materials on engagement of persons who are profoundly, multiply handicapped. *Education and Training in Mental Retardation, 23*(3), 228–237.

Reese, G. M., & Snell, M. E. (1991). Putting on and removing coats and jackets: The acquisition and maintenance of skills by children with severe multiple disabilities. *Education and Training in Mental Retardation, 26*, 398–410.

Reichow, B., & Wolery, M. (2011). Comparison of progressive prompt delay with and without instructive feedback. *Journal of Applied Behavior Analysis, 44*, 327–340.

Schuster, J. W., Morse, T. E., Ault, M. J., Doyle, P. M., Crawford, M. R., & Wolery, M. (1998). Constant time delay with chained tasks: A review of the literature. *Education and Treatment of Children, 21*, 74–106.

Sigafoos, J., O'Reilly, M., Seely-York, S., & Edrisinha, C. (2004). Teaching students with developmental disabilities to locate their AAC device. *Research in Developmental Disabilities, 25*(4), 371–383.

Sisson, L. A., & Dixon, M. J. (1986). Improving mealtime behaviors of a multi-handicapped child using behavior therapy techniques. *Journal of Visual Impairment & Blindness, 80*, 855–858.

Sisson, L. A., Kitwein, M. L., & Van Hasselt, V. B. (1988). A graduated guidance procedure for teaching self-dressing skills to multihandicapped children. *Research in Developmental Disabilities, 9*, 419–432.

Sonksen, P., Levitt, S., & Kitzinger, M. (1984). Identification of constraints acting on motor development in young visually disabled children and principles of remediation. *Child: Care, Health and Development, 10*, 273–286.

Spencer, R. A., Head, D. N., Van Dusen Pysh, M., & Chalfant, J. C. (1997). Response patterns of children with visual impairments on measures of internalized self-responsibility. *Review: Rehabilitation and Education for Blindness and Visual Impairment, 29*, 121–127.

Strawbridge, L. A., Drnach, M., Sisson, L. A., & Van Hasselt, V. B. (1989). Behavior therapy combined with physical therapy to promote walker use by a child with multiple handicaps. *Behavior Therapy, 24*, 239–247.

Striefel, S., Wetherby, B., & Karlan, G. (1978). Developing generalized instruction-following behavior in severely retarded people. *Monograph of the American Association of Mental Deficiency*, *3*, 267–326.

Taber-Doughty, T. (2005). Considering student choice when selecting instructional strategies: A comparison of three prompting systems. *Research in Developmental Disabilities*, *26*(5), 411–432.

Taras, M., Matson, J. L., & Felps, J. N. (1993). Using independence training to teach independent living skills to children and young men with visual impairments. *Behavior Modification*, *17*, 189–208.

Taylor, R. G. (1987). Teaching a severely handicapped deaf-blind young woman to prepare breakfast foods. *Journal of Visual Impairment & Blindness*, *81*, 67–69.

Taylor, P., Collins, B. C., & Schuster, J. W. (2002). Teaching laundry skills to high school students with disabilities: Generalization of targeted skills and nontargeted information. *Education and Training in Developmental Disabilities*, *37*(2), 172–183.

Taylor, J. C., McKelvey, J. L., & Sisson, L. A. (1993). Community-referenced leisure skill clusters for adolescents with multiple disabilities. *Journal of Behavioral Education*, *3*, 363–386.

Tekin-Iftar, E., Kircaali-Iftar, G., Birkan, B., Uysal, A., Yildirim, S., & Kurt, O. (2001). Using a constant delay to tech leisure skills to children with developmental disabilities. *Mexican Journal of Behavior Analysis*, *27*, 337–362.

Tekin-Iftar, E., Acar, G., & Kurt, O. (2003). The effects of simultaneous prompting on teaching expressive identification of objects: An instructive feedback study. *International Journal of Disability, Development and Education*, *50*(2), 149–167.

Touchette, P. E. (1971). Transfer of stimulus control: Measuring the moment of transfer. *Journal of the Experimental Analysis of Behavior*, *15*, 347–354.

Trask-Tyler, S. A., Grossi, T. A., & Heward, W. L. (1994). Teaching young adults with developmental disabilities and visual impairments to use tape-recorded recipes: Acquisition, generalization, and maintenance of cooking skills. *Journal of Behavioral Education*, *4*, 283–311.

U.S. Department of Commerce, Economics and Statistics Administration, U.S. Census Bureau. (2011). *School-aged children with disabilities in U.S. metropolitan statistical Areas: 2010 American community survey briefs*, (U.S. Census Publication No. ACSBR/11-12). Retrieved from http://www.census.gov/prod/2013pubs/acsbr11-12.pdf.

Venn, M. L., Wolery, M., Werts, M. G., Morris, A., DeCesare, L. D., & Cuffs, M. S. (1993). Embedding instruction in art activities to teach preschoolers with disabilities to imitate their peers. *Early Childhood Research Quarterly*, *8*, 277–294.

Walker, G. (2008). Constant and progressive time delay procedures for teaching children with autism: A literature review. *Journal of Autism and Developmental Disorders*, *38*, 261–275.

Warren, D. H., & Hatton, D. D. (2003). Cognitive development of children with visual impairments. In *Part II: Child neuropsychology: Vol. 8. Handbook of neuropsychology* (pp. 439–458) (2nd ed.) New York, NY: Elsevier.

Waugh, R. E., Alberto, P. A., & Fredrick, L. D. (2011). Simultaneous prompting: An instructional strategy for skill acquisition. *Education and Training in Autism and Developmental Disabilities*, *46*, 528–543.

West, E. A., & Billingsly, F. (2005). Improving the system of least prompts: A comparison of procedural variations. *Educationand Training in Developmental Disabilities*, *40*(2), 131–144.

Wolery, M., Busick, M., Reichow, B., & Barton, E. E. (2010). Comparison of overlap methods for quantitatively synthesizing single-subject data. *Journal of Special Education*, *44*, 18–28.

Wolery, M., & Gast, D. L. (1984). Effective and efficient procedures for the transfer of stimulus control. *Topics in Early Childhood Special Education*, *4*(3), 52–77.

Wolery, M., & Gast, D. L. (2000). Classroom research for young children with disabilities: Assumptions that guided the conduct of research. *Topics in Early Childhood Special Education*, *20*(1), 49–55.

Wolery, M., Holcombe, A., Cybriwsky, C., Doyle, P. M., Schuster, J. W., Ault, M. J., et al. (1992). Constant time delay with discrete responses: A review of effectiveness and demographic, procedural, and methodological parameters. *Research in Developmental Disabilities*, *13*, 239–266.

Wolery, M., & Schuster, J. W. (1997). Instructional methods with students who have significant disabilities. *The Journal of Special Education*, *31*, 61–83.

Wright, T., & Wolery, M. (2012). Evaluating the effectiveness of roadside instruction in teaching youth with visual impairments street crossings. *The Journal of Special Education* http://dx.doi.org/10.1177/0022466912453770.

Yilmaz, I., Birkan, B., Konukman, F., & Erkan, M. (2005). Using a constant time delay procedure to teach aquatic play skills to children with autism. *Education and Training in Developmental Disabilities*, *40*(2), 171–182.

CHAPTER FOUR

Communication Development of Children with Visual Impairment and Deafblindness: A Synthesis of Intervention Research

Amy T. Parker[*,1], **Sarah E. Ivy**[†]
[*]Western Oregon University, Monmouth, Oregon, USA
[†]Department of Special Education, Vanderbilt University, Nashville, Tennessee, USA
[1]Corresponding author: e-mail address: parkera@wou.edu

Contents

International Review of Research in Developmental Disabilities, Volume 46
ISSN 2211-6095
http://dx.doi.org/10.1016/B978-0-12-420039-5.00006-X

Abstract

This chapter reports the results of a synthesis of research on educational interventions to improve early and emergent communication skills, both expressive and receptive, for children and youth with visual impairment and deafblindness, birth to 22 years of age. A search of electronic databases and recently published systematic reviews returned 34 articles published from January 2000 to August 2013 meeting inclusion criteria. Intervention studies focused primarily on communication partner training ($n=12$), microswitch use ($n=10$), or object symbol use ($n=7$). The majority of researchers used experimental single-subject methods ($n=21$, 62%) to examine efficacy of intervention to improve expressive communication skill. Results are discussed in terms of student characteristics, settings for intervention, critical procedural parameters, and targeted communication skills. Key conclusions for practitioners and researchers include individualizing interventions based on student preferences, ensuring the accessibility of materials and communication partners, implementing interventions within meaningful and naturalistic daily routines, training communication partners to be responsive, and designing and implementing high-quality research to identify evidence-based practices for communication interventions.

1. INTRODUCTION

The human body is a vehicle for multiple types of behavioral expressions. Some of these behaviors, in the form of traditional speech or through nonverbal expressions, may be interpreted as communicative. This "liberal" definition of communication has emerged over the past 40 years of research with infants (Bates, Benigni, Bretherton, Camaioni, & Volterra, 1977; Rowland, 1990). From birth, it is through vision, hearing, touch, taste, smell, and proprioception that environmental connections, organization, conceptual relationships, and the foundation for communication with others are formed (van Dijk, 1969; Werner & Kaplan, 1964). Before infants speak, the process of developing intentional communicative behaviors is practiced in everyday environments during daily routines. Prelinguistic behaviors, such as vocalizations, eye gaze, gestures, body movements, postural changes, and even breathing patterns, may function as communication for infants and their caregivers. The expansion of behaviors that may be recognized as communicative provides a broader platform for intervention for individuals who have severe communication impairments and has helped shape the field of augmentative and alternative communication (AAC; Johnston, Reichle, & Evans, 2004). AAC refers to forms of communication and resources that enhance or supplement communication, speech, and writing (U.S. Department of Education: Office of Special Education and Rehabilitative Services, 2001). AAC, considered one type of assistive technology under the Individuals with Disabilities Education Improvement Act of 2004, has been described as promoting social relationships, participation, self-determination, increased employment, and independence.

1.1. Expanded Definition of Communication

By acknowledging that simple, nonspeech behaviors may be interpreted by partners as requests, greetings, protests, or even comments, a framework for viewing all persons as communicators, even those with nascent communicative abilities, has emerged (Hourcade, Pilotte, West, & Parette, 2004). Further, as a part of the evolution of the field of AAC, many researchers, teachers, and therapists recognize that supporting individuals with severe communication impairments involves supporting or intervening within a type of ecosystem (Sigafoos, 2005). Bruce, Godbold, and Naponelli-Gold (2004) described this system as having four dynamically interrelated components: forms (mode or vehicle for expressing and receiving), content (messages), functions (purpose of the behavior), and context (the situation and elements in the environment). Inherent in this broad definition of a communication system is the role of the communication partner as interpreter and active participant (Rowland, 1990). Communication, then, not only involves the expressed behavioral form of the person sending a message, verbal or nonverbal, but also requires a receiver who assigns meaning to perceived communicative attempts (Grove, Bunning, Porter, & Olsson, 1999). For communication to be successful, meaning must be shared.

1.2. Challenges in Development of Communication for Children with VI and DB

The simple act of engaging in a communicative exchange may be tremendously challenging for young children with multiple disabilities, particularly those with concomitant visual impairment (VI) or deafblindness (DB) (Bruce, 2005; Schweigert, 1989). Qualitative descriptive studies conducted over the past several decades describe the disrupted communication patterns and the barriers to communication that exist between young children with VI or DB and their caregivers (Rowland, 1990; Vervloed, van Dijk,

Knoors, & van Dijk, 2006). Because children with vision and multiple disabilities (VIMD) may not be able to access environmental cues that support communication, they can develop limited communicative repertoires or highly idiosyncratic ways of communicating, which may be unrecognizable to communication partners (Siegel-Causey & Downing, 1987; Vervloed et al., 2006). Researchers who have measured communication exchanges between children with VIMD or DB and their partners reported that even individuals with intimate knowledge of the child may not recognize, support, or respond to the child's communication attempts/behaviors, thereby decreasing opportunities for building communication skills (Bruce et al., 2004; Vervloed et al., 2006). Other theorists and practitioners have described the phenomenon known as "learned helplessness" that occurs when children do not recognize that their communication attempts produce consistent responses from partners (Schweigert, 1989; Seligman, 1975). In behavioral approaches to intervention, this lack of response from partners may be categorized as "extinguishing" nascent communication behaviors rather than supporting them.

Communication theorists, practitioners, and researchers have described vision as an organizing and contextualizing sense that provides access to environmental cues and supports for engaging in communication (Anderson, Dunlea, & Kekelis, 1984; House & Davidson, 2000; Pérez-Pereira & Castro, 1997; Warren & Hatton, 2003). Vision also provides access to an array of topics and the awareness that a partner is available and interested in communication (Bruce, 2005). Vision also serves as a memory device and a prompt in the environment, drawing people into conversations, even of the most basic type, about what is immediately before them. Importantly, vision supports the development of cause and effect, but it also is a way to share interest and attention, often the foundation of emergent communication behaviors (House & Davidson, 2000; Warren & Hatton, 2003).

For children with DB, there are compounded risks for not developing a reliable communication system. Unlike other children with VIMD, individuals with DB may not be able to access speech sounds or may have a completely altered sense of speech sounds, even with amplification (Broadston, 2003). Although it may be impossible to determine the impact of VI upon students with complex, concomitant disabilities, VI and the more profound disability of DB may impede initially developing communication skills (Rowland & Stremel-Campbell, 1987; Warren, 1994). Young children with VI, particularly those with multiple disabilities or

DB, have more subtle ways of responding to their communicative partners, and without intervention, communication partners often fail to recognize these responses as opportunities to communicate (Dote-Kwan & Hughes, 1994; Rowland, 1990; Warren & Hatton, 2003).

1.3. Challenges of Research with Children with VI and DB

Central to any discussion of communication interventions with children who have VI or DB is an understanding of the challenges in conducting research with low incidence and highly heterogeneous populations (Ronnberg & Borg, 2001). According to the American Printing House for the Blind's annual report, there are approximately 58,939 school-aged students who are identified as legally blind in the United States (APH, 2011). Because the educational definition of children with VI includes children with *low vision* and *functional blindness* who may not meet the criteria for legal blindness, the actual number of children with VI served in schools is estimated to be 120,000 (Mason & Davidson, 2001).

Children with VI are highly diverse, and children with DB are even more so (Ferrell, 2000; Hatton, Ivy, & Boyer, 2013; Schalock & Bull, 2012). Within Ferrell and colleagues' 1998 sample of 202 infants and toddlers with VI, 60% were identified as having additional disabilities. Of this subset of children, approximately 90 different coexisting impairments were documented (Ferrell). Additionally, over 65% of a recent sample of young children with severe uncorrectable VI, referred for early intervention services in the United States between 2005 and 2011, had additional disabilities or developmental delay (Hatton et al., 2013). Currently, there are just over 10,000 children with DB who are counted in the federally mandated child count (Schalock & Bull, 2012). Within this disability category, there is great heterogeneity in terms of degrees of sensory loss, for both vision and hearing, as well as the types of additional disabilities that are present. According to Schalock and Bull, approximately 90% of all infants who are congenitally deafblind have additional disabilities. Further, in an examination of studies that included students with intellectual disabilities and multiple disabilities, concomitant VI was found to be 200 times more likely than in the typical population (Warburg, 2008). Medical causes of intellectual disabilities or various syndromes may also be associated with significant VI and DB (Schalock & Bull, 2012). Although VI and DB are often found in students with multiple disabilities, local service providers may not identify the sensory losses at an early age, and they may not recognize the impact of sensory loss in developing reliable communication systems (Rowland, 2009). Despite the heterogeneity of students who have VI or DB, significant communication delays and impairments within this diverse group are a common denominator (Bruce, 2005; Ronnberg & Borg, 2001).

1.4. Previous Reviews

Over the years, communication interventions for students with VIMD, the majority of which may be categorized as AAC interventions, have been implemented in a number of environments (Parker, Grimmett, & Summers, 2008; Sigafoos et al., 2008). In Sigafoos and colleagues' review of studies that included individuals who are deafblind, only research that could be classified as experimental was included. Of the 17 intervention studies analyzed, only six were found to have conclusive evidence of effectiveness. Parker and her collaborators sought to identify communication interventions that included participants under the age of 22 with VIMD in their review of single-subject design studies.

From each of these reviews, some unique approaches for intervention have been identified for individuals who have VIMD or DB. For example, many AAC interventions used with persons who have VIMD have required some adaptations to compensate for participants' reduced or absent visual acuity (Parker et al., 2008; Sigafoos et al., 2008). Examples of these adaptations include backlighting a microswitch (Singh et al., 2003), adding textures to the surfaces of a calling device (Mathy-Laikko et al., 1989; Schweigert, 1989), using three-dimensional parts of objects as referents for actual items (Rowland & Schweigert, 2000; Trief, 2007; Turnell & Carter, 1994), employing tactual sign language as a multicomponent partner intervention (Janssen, Riksen-Walraven, & van Dijk, 2003), and making communication boards with large print, bold line drawings, and words (Heller & Allgood, 1996). Each review stressed the importance of consistent support from communication partners (Parker et al., 2008) and the need to have good preintervention assessment data on participants' vision and hearing (Sigafoos et al., 2008).

Although these reviews are helpful in examining communication interventions for this unique population, authors view the literature with a very specific methodological lens. Inclusion criteria for each review emphasized broad experimental approaches (Sigafoos et al., 2008) or single-subject designs (Parker et al., 2008). Sigafoos and colleagues analyzed the available information on communication skills gained through the interventions, but only participants with DB were included. Parker and colleagues categorized approaches into types of interventions, but overall participant gains were not reported.

1.5. Need for High-Quality Efficacious Interventions

In recognition of the complex challenges in conducting socially relevant, educationally based intervention research, the Council for Exceptional Children (CEC) has acknowledged the importance of different types of research methods to validate various intervention approaches (Odom et al., 2005). Further, the American Speech, Language, and Hearing Association's (ASHA, 2005) statement on evidence-based practices includes the integration of high-quality clinical expertise with scientific evidence to undergird communication intervention practices. Further, several federally funded projects that emphasize building communication skills with this population have emphasized the need for practically oriented, translational research that includes high-quality, individualized intervention strategies that are applied in everyday contexts (Chen, Klein, & Haney, 2000).

The most compelling reason to explore recent intervention research is the need for students to have access to more reliable and recognizable ways to communicate across environments. Teachers, parents, and related service providers have an immediate and practical need for information that can assist them in finding appropriate AAC systems for students with VIMD and DB (Millar, Light, & Schlosser, 2006; Parker et al., 2008). Therefore, the purpose of this chapter is to describe interventions that have been shown to increase early communication in students with VI, less than 22 years old. To achieve the goal of identifying effective intervention practices for developing communication, we identified and analyzed research studies published in peer-reviewed journals between 2000 and August 2013 to describe (a) the outcomes (dependent variables), with particular attention to communicative form and function; (b) the intervention procedures, including settings and communication partners; (c) the characteristics of the students in the sample; and (d) the types of research methods used. By systematically examining this information, we will also offer (e) implications for professionals and families and (f) priorities for future research.

2. METHOD

To identify the studies for this review, we searched Eric, Psycharticles, PsychInfo, and Proquest Central Education databases, combining the following search terms: *intervention, support, strategy, approach,* or *teaching*; with *elementary, young, develop**, *child**, *emergent, preschool,* or *early*; with *communicat**, *participat**, *interact**, *alertness, awareness, biobehavioral/states, language, literacy, social/exchang**, *social/interact**, *AAC, augmentative/communication, augmentative/and/alternative, alternative/communication,* or *engagement*; with *vis**/*impair**, *vis**/*loss, vis**/*disab**, *multipl**/*disab**, *deafblind**, *deaf-blind**, *deaf/blind, dual**/*impair**, *sensory/impair**, *multi-handicap**, or *vis**/*handicap**. These terms were also combined with terms of diagnoses taken from the Babies Count Survey ($N=32$; Hatton et al., 2013) and the National Deaf-Blind Child Count ($N=79$; Schalock & Bull, 2012). This search returned 318 articles published in peer-reviewed journals between 2000 and August 2013.

To be included in this review, studies had to (a) be a research study, (b) evaluate a communication intervention, (c) include at least one participant with VI or DB who was less than 22 years old, and (d) report student data on an early communicative outcome attributed to the intervention. For this review, research study meant an article in a peer-reviewed journal that had a specific research question or goal and that reported data on participants who met our inclusion criteria. *Communication intervention* was loosely defined as something being done with someone to increase the ability to convey feelings, thoughts, wants, or needs to another person or to receive and respond to such messages from another person. Medical interventions (e.g., cochlear implantation), surveys, or other perception ratings, program evaluations, and purely descriptive studies were excluded. VI included any reference to an uncorrectable impairment of the eye or brain that negatively affected the students' vision. DB included any descriptors that indicated a combination of vision and hearing impairment. *Early communicative outcomes* were defined as dependent variables that reflect participants' emerging ability to understand or convey a message to another person for the purpose(s) of, but not limited to, rejecting, requesting, achieving social closeness, or commenting.

For article selection, we reviewed the results from our literature search three times. First, we reviewed only titles and abstracts and excluded reviews, articles with no participants with disabilities, and those that were not communication-related. We also conducted an ancestral search of systematic reviews returned from our database search. During the second review of full text articles, we excluded nonintervention studies and studies that did not include participants with VI or did not target a communication skill. For the third review of full text articles, we narrowed our inclusion criteria to participants less than 22 years old and excluded studies that did not include participant interactions with communication partners. For example, we excluded articles that involved participants learning to control environmental stimuli, except in cases in which speech articulation or receptive understanding of speech was targeted. Additionally, we included several studies that focused on training communication partners to be more responsive if child/student data were reported.

To estimate the reliability of article selection, an independent rater reviewed a subsample of the studies to apply the inclusion/exclusion criteria during each of the three reviews. During the first pass, the first and second rater trained to 90% agreement, and then, the second rater reviewed 20% of the first 20 articles. If agreement was above 90%, the raters continued to the next set of 20 articles. If 90% agreement was not met, discrepancy discussions were conducted for disagreements, and then, the second rater reviewed all 20 articles in that set. Agreement was checked for the entire set, and discrepancy discussions were conducted for disagreements. During discrepancy discussions, the decision to include an article was made by consensus. This process continued until all studies were reviewed. For the second and third pass, after training to 100%, one rater reviewed 100% of the studies and the second rater reviewed 20% of the total sample. Interrater agreement for article selection was calculated across all three reviews using the formula: number of agreements divided by the number of agreements plus disagreements multiplied by 100. Overall agreement was 88% for article selection.

2.1. Analysis

Studies meeting final inclusion criteria were analyzed to describe the following: research methods, participants, intervention components, communication partners, settings for intervention, form and function of the dependent measure, and the study results. The CEC's classification of research methodologies was used to describe research methods (Odom et al., 2004). Descriptions of participants, intervention components, and communication partners were taken verbatim from studies. Settings for intervention were categorized at as specific level as possible based on the information reported. For example, if researchers did not specify that intervention had been provided in a classroom, we reported the location at the school level (special education school, residential school, or inclusive school). Description of the form and function of the dependent measure was taken verbatim from studies or classified according to the forms and functions as defined in

Rowland (2009). Included in the description of form is the means of access. For example, if a microswitch device was used for expressive communication, the part of the body the student used to access the device was included.

Finally, we report when a functional relation was demonstrated in single-subject studies and estimated the size of positive effects according to the following scale: If a therapeutic change in the dependent variable was reported in the range of approximately 10–40% increase from baseline performance, small effects were noted; a change in the range of approximately 41–70% was considered moderately effective, and above approximately 71% was considered largely effective. In the case of 0% responding in baseline, the percentage change from baseline to intervention was based on performance at the 1% level in baseline. Mixed results were reported when the intervention was not effective for some participants or some behaviors. Visual analysis was used to determine a functional relation in single-subject design studies. A functional relation was defined as a demonstration of a stable baseline or countertherapeutic trend followed by change in level or trend of the dependent variable that covaried with manipulation of the independent variable, within the context of an experimental design recognized by CEC (Horner, Carr, Halle, Odom, & Wolery, 2005) with at least three successful replications of effect. Level was defined as an abrupt change between the last data point in baseline and one of the first three data points during intervention. Trend was defined as change in the slope of the dependent measure from baseline to intervention.

3. RESULTS

Thirty-four studies met final inclusionary criteria for analysis. Twenty-one of the 34 employed experimental single-subject design approaches. In another seven studies, repeated measurement of behavior was completed in the context of one or more AB designs. In three studies, assessment of a single group of participants was only conducted pre- and postintervention. Another three studies used qualitative methods. There were 148 participants with VI or DB ranging in age from infancy to 22 years in the studies we reviewed.

Descriptors of the conditions associated with VI or DB were wide-ranging, including general references to reduced vision or hearing abilities, as well as specific etiologies. A sample of descriptors used across studies included optic nerve atrophy, septo-optic dysplasia, Leber's congenital amaurosis, congenital rubella syndrome, Cornelia de Lange syndrome,

Zellweger's syndrome, retinopathy of prematurity, Robinow syndrome, CHARGE syndrome, Pierre Robin syndrome, cytomegalovirus, perinatal hypoxia, and sclerocornea. Tables 4.1–4.6 denote the stated conditions included in the published reports. Specific descriptors of participants with moderate or profound intellectual impairment were included in 22 studies. For the remaining 12 studies, descriptors used to denote the participants' communication or adaptive functioning within various settings indicated that participants probably had intellectual disabilities. One study included an infant participant for whom cognitive testing was not appropriate (Lappin, 2005). Twenty studies also included information about motor impairments.

Intervention was provided in special education settings in 29 of 34 studies. Additionally, intervention was provided in inclusive educational programs in three studies and home settings in three studies. In 19 studies, a researcher was the primary communication partner. In 12 studies, partners were described as teachers, paraprofessionals, or residential caregivers. In two studies each, parents or peers served as communication partners. The communication forms and functions targeted for instruction are documented in Tables 4.1–4.6.

Although the types of intervention approaches varied, patterns emerged. These general categories include microswitch or computer software technologies to support communication ($n=10$); multicomponent communication partner training ($n=12$); object symbol studies, including adapted Picture Exchange Communication System (PECS) interventions ($n=7$); literacy- or narrative-based approaches that specifically measured changes in communication behaviors ($n=4$); the use of wait time ($n=1$); and the use of adapted prelinguistic milieu training ($n=1$).

3.1. Microswitch and Technological Supports

This category comprised 10 studies in which microswitch devices or software systems were used to enhance students' ability to participate in an interaction with a partner or to enhance participants' use of their own voice to make requests or choices. Most researchers emphasized the functional aspects of expressive behaviors to increase the participants' responsiveness and control of the interaction. All researchers employed experimental single-subject research methodology, except in two studies in which AB designs were used, to measure the efficacy of the intervention approaches.

Table 4.1 Summary of Results for Studies of Microswitch and Technological Supports

Study	VI or DB Participants	ID	Intervention Components	Partner	Context	Research Design	Outcomes: Form	Outcomes: Function	Outcomes: Results
Cosbey and Johnston (2006)	*N*=2, 4–6 yrs agenesis of corpus callosum, vision difficulties, wheelchair	ID (severity not reported)	Microswitch, naturalistic play routines, teacher support	Peers without disabilities	Inclusive classroom (center time, accessing toys, social interaction with peers)	MP—participants	Activating switch	Requests	Functional relation, large effects
Lancioni, O'Reilly, et al. (2009)	*N*=10, 5–18 yrs MD blindness, prematurity, hypoxia	Severe–profound	Microswitch and VOCA responses, preference assessment	Researcher	SPED center	MP—behaviors	Activating with arm, hand, head turning, vocalization	Requests for social interaction	Functional relation, moderate effects
Lancioni, Singh, et al. (2009)	*N*=5, 9–22 yrs perinatal hypoxia, cerebral hemorrhaging, blindness, residual vision, wheelchair	Severe–profound	Microswitch, throat microphone, preference assessment	Researcher	SPED center	MP—behaviors	Activating with throat	Social, choice making	Functional relation, small–moderate effects

Lancioni et al. (2008)	*N*=3, 10–15 yrs blind, low vision, wheelchair	Severe–profound	Microswitch, contingent reinforcement, preference assessment	Researcher	SPED center	MP—behaviors	Activating with trunk/head, hand, foot movements	Requests, social contact	Positive, small effects
Lancioni et al. (2006)	*N*=2, 9–12 yrs blindness, prenatal hypoxia	Severe–profound	Dragon dictate, throat microphone, preference assessment	Researcher	SPED center	MP—behaviors	Vocal utterances	Choice making	Positive, moderate–large effects
Lancioni et al. (2005)	*N*=1, 18 yrs wheelchair	Severe	Dragon dictate, computer, microphone, preference assessment	Researcher	SPED center	MP—behaviors	Vocal utterances	Improved speech, requests	Functional relation, small–moderate effects
Lancioni, Singh, O'Reilly, et al. (2004)	*N*=2, 18–22 yrs wheelchair	Profound	Dragon dictate, computer, microphone, modeling, prompts, preference assessment	Researcher	SPED center	MP—behaviors	Vocal utterances	Improved speech, requests	Functional relation, moderate effects

Continued

Table 4.1 Summary of Results for Studies of Microswitch and Technological Supports—cont'd

Study	VI or DB Participants	ID	Intervention Components	Partner	Context	Research Design	Outcomes: Form	Outcomes: Function	Outcomes: Results
Lancioni, Singh, O Reilly, Oliva, and Montironi (2004)	*N*=2, 18–20 yrs congenital hydrocephalus prematurity, perinatal hypoxia, blind wheelchair	Profound	Dragon dictate with computer, microphone, preference assessment	Researcher	SPED center	AB design, pre/post	Word utterances interpreted by speech recognition software	Improved speech, requests	Positive, moderate effects
Lancioni, Singh, O'Reilly, and Oliva (2003)	*N*=2, 17–19 yrs blind, spastic tetraparesis	Severe–profound	Speech recognition software as microswitch, preference assessment	Researcher	SPED center	MP—behaviors	Increasing utterances—shaping spoken utterances	Improved speech, requests	Functional relation, moderate–large effects
Lancioni, Singh, O'Reilly, Oliva, Dardanelli, et al. (2003)	*N*=2, 14–18 yrs ONH, blindness hydrocephalus, wheelchair	Severe–profound	Microswitch activated by vocalization—prompts, physical and verbal, preference assessment	Researcher	SPED center	AB design, pre/post	Word cues to activate speech output device	Improved speech, requests	Positive, moderate effects, improved scores on word association test

Table 4.2 Summary of Results for Studies of Multicomponent Communication Partner Training

Study	VI or DB Participants	ID	Intervention Components	Partner	Context	Research Design	Outcomes: Form	Outcomes: Function	Outcomes: Results
Grisham-Brown et al. (2000)	$N=4$, 2–4 yrs DB, wheelchair	Severe	Training paras via video and coaching, embedded instruction, response prompting	Paraeducators	Inclusive Preschool (activities of daily living, center time)	MP—participants	Switches, using sign language within routines	Requests, following directions	Functional relation, moderate–large effects for some behaviors
Lappin (2005)	$N=1$, infant ROP	Not reported	Infant massage, parent training	Mother	Home (free play, interactive games)	Case study	Cooing orienting smiling vocalization	Social/turn-taking	Increased interaction and positivity
Horrocks and Morgan (2011)	$N=3$, 6 yrs MD, strabismus, legally blind	Severe–profound	1:1 Teacher training; video samples; role play and coaching, preference assessment, response prompting	Teachers	SPED classroom (academic instruction)	MP—participants	Touching items, eye gaze, push items away	Requests, rejecting, social	Functional relation, moderate effects

Continued

Table 4.2 Summary of Results for Studies of Multicomponent Communication Partner Training—cont'd

Study	VI or DB Participants	ID	Intervention Components	Partner	Context	Research Design	Outcomes: Form	Outcomes: Function	Outcomes: Results
Damen et al. (2011)	*N*=3, 13–19 yrs blind and partial sight	Severe–profound	Diagnostic intervention model (DIM)	Residential caregivers	Special residential school	Multiple AB designs, pre/post	Orienting, laughing, reaching hands or arms to caregivers	Rejecting, requests, social, informing/asking	Positive but no significant differences for student behavior
Janssen et al. (2012)	*N*=6, 3–19 yrs DB, CRS, Leber's, Cornelia de Lange, Zellweger's	ID (severity not reported)	DIM	Teachers	Special residential school	Multiple AB designs	Gestures, touching objects, drawings, tangible symbols	Rejecting, requests, social, informing/asking	Positive for sustained interactions
Janssen et al. (2011)	*N*=1, 5 yrs, DB	ID (severity not reported)	DIM	Educators	Special residential school	MB—behaviors	Drawings, gestures, decrease SIB, positive exchanges	Rejecting, requests, social, informing/asking	Functional relation, positive for some behaviors, decrease negative behaviors

Janssen et al. (2010)	*N*=2, 3 yrs DB	ID (severity not reported)	DIM	Mothers	Home	AB design	Touching objects, smiling vocalization; interaction games	Rejecting, requests, social, informing/asking	Positive, moderate effects for multiple behaviors
Janssen, Riksen-Walraven, Van Dijk, Ruijssenaars, and Vlaskamp (2007)	*N*=2, 16–19 yrs DB CRS	ID (severity not reported)	DIM	Teacher caregivers	Special residential school	AB design	Calendar system touch; tactile signs; tactile symbols	Rejecting, requests, social, informing/asking	Positive, mixed results
Janssen, Riksen-Walraven, and van Dijk (2006)	*N*=1, 10 yrs, glaucoma, CRS	ID (severity not reported)	DIM	Teacher (1)	Special residential school	AB design	Tangible symbols (comm. book)	Rejecting, requests, social, comment informing/asking	Positive, moderate effects for multiple behaviors
Janssen, Riksen-Walraven, and van Dijk (2004)	*N*=4, 7–11 yrs DB, MD	ID (severity not reported)	DIM	Educators (16)	Special residential school	MB—participants	Drawings, gestures, objects spoken words,	Rejecting, requests, social, informing/asking	Functional relation, small effects

Continued

Table 4.2 Summary of Results for Studies of Multicomponent Communication Partner Training—cont'd

Study	VI or DB Participants	ID	Intervention Components	Partner	Context	Research Design	Outcomes		
							Form	Function	Results
							positive interactions		
Janssen et al. (2003)	$N=6$, 3–19 yrs DB, MD	ID (severity not reported)	DIM	Teachers and residential caregivers (14)	Special residential school	MB—participants, pre/post	Vocalizing, signs, words, positive exchanges, tangible symbols	Rejecting, requests, social, informing/asking	Positive, large effects
Janssen et al. (2002)	$N=4$, 6–9 yrs DB, MD	ID (severity not reported)	DIM	Teachers and residential caregivers (14)	Special residential school	MB—participants	Vocalizing, signs, words, positive interactions	Rejecting, requests, social, informing/asking	Functional relation, small effects

Table 4.3 Summary of Results for Studies of Tangible Symbols, Object Symbols, and Adapted PECS Interventions

Study	VI or DB Participants	ID	Intervention Components	Partner	Context	Research Design	Outcomes: Form	Function	Results
Ali, MacFarland, and Umbreit (2011)	$N=4$, 7–14 yrs ONH, nystagmus	Moderate	Adapted PECS, preference assessment	Researcher	SPED classroom (leisure)	MB—participants	Exchange of tangible symbols	Requests	Functional relation for Phases 1–3
Ivy et al. (in press)	$N=4$, 5–11 yrs ONH, SOD, wheelchair	Severe	Adapted PECS, preference assessment	Researcher	SPED classroom (leisure)	MP—participants	Exchange of tangible symbols	Requests	Functional relation for Phase 1, mixed results for Phase 2
Lund and Troha (2008)	N = 3, 12–17 yrs ONH, autism	Severe–profound	Adapted PECS, preference assessment	Researcher	SPED classroom (leisure)	MB—participants	Exchange of tangible symbols	Requests	Functional relation for Phase 1
Parker, Devender, Davidson, and Liu-Gitz (2009)	$N=1$, 7 yrs Leber's	Severe	Adapted PECS, preference assessment	Researcher	SPED classroom (snack time, free play)	Changing criterion	Exchange of tangible symbols	Requests	Positive for Phases 1–4
Rowland and	$N=23$, 3–18 yrs DB	ID (severity	Tangible symbols (communication book/strip),	Researcher	Inclusive and SPED classroom	Pre/post	Touching, pointing, or	Requests, social,	80% Learned symbols and improved scores

Continued

Table 4.3 Summary of Results for Studies of Tangible Symbols, Object Symbols, and Adapted PECS Interventions—cont'd

Study	VI or DB Participants	ID	Intervention Components	Partner	Context	Research Design	Outcomes: Form	Outcomes: Function	Outcomes: Results
Schweigert (2000)		not reported)	preference assessment				extending objects	informing/ asking	on communication matrix
Trief (2007)	$N=25$, 4–16 yrs DB	Moderate–severe	Tangible symbols (schedule)	Teachers	SPED school (transitions)	Pre/post	Touching or pointing to tangible symbols	Answering questions/ informing	60% Learned 1–28 symbols
Harding, Lindsay, O'Brien, Dipper, and Wright (2011)	$N=1$, 6 yrs wheelchair	Severe–profound	Objects of reference, multiple exemplars, hand-over-hand prompting, wait time, simplified speech and gestures	Researcher, teacher, and SLP, general to peers	SPED classroom (music, free play, lunch)	Pre/post	Touching objects	Choice making	Improved scores on pragmatics profile of everyday communication skills ($p<0.05$)

Table 4.4 Summary of Results for Studies of Narrative- or Literacy-Based Interventions

Study	VI or DB Participants	ID	Intervention Components	Partner	Context	Research Design	Outcomes: Form	Outcomes: Function	Outcomes: Results
Browder et al. (2011)	$N=1$, 9 yrs legally blind, wheelchair	Severe	Adapted books, objects, pictures, interactive shared storybooks, systematic prompting	Teachers	SPED classroom	MP—behaviors	Touching objects, printed names, and books	Answer questions/inform	Functional relation, moderate–large effects
Mims et al. (2009)	$N=2$, 6–9 yrs CVI, MD, severe myopia, microcephaly, CP, wheelchair	Severe–profound	Adapted books, objects, repeated story lines, least to most prompts, opportunities to respond	Researcher, general to peers	SPED school and classroom (cafeteria, literacy activity)	MP—behaviors	Touching objects	Answer questions/inform	Functional relation, moderate–large effects
Park (2004)	$N=5$, school-aged (specific ages not reported) DB	Profound	Adapted switches, objects, repeated story lines, opportunities to respond	Researcher, peers with disabilities, teachers	SPED classroom (dramatic reading in a small group with props)	Qualitative (encounter awareness framework)	Switch activations, orienting, engaging, turn-taking, joint attention	Inform (perform part in dramatic reading narrative)	Positive effects for multiple behaviors

Table 4.5 Summary of Results for Studies of Wait Time

Study	VI or DB Participants	ID	Intervention Components	Partner	Context	Research Design	Outcomes		
							Form	Function	Results
Johnson and Parker (2013)	$N=2$, <11 yrs DB, MD, CVI, optic nerve scarring, SOD	ID (severity not reported)	Wait time, preference assessment	Researcher	Home (free play)	Alternating treatment	Receptive verbal messages	Answer questions/ inform	Positive results for increasing wait times

Table 4.6 Summary of Results for Studies of Adapted Prelinguistic Milieu Teaching

							Outcomes		
Study	VI or DB Participants	ID	Intervention Components	Partner	Context	Research Design	Form	Function	Results
Brady and Bashinski (2008)	$N=9$, 3–7 yrs DB, orthotic impairments	Severe	Adapted RPMT, hand-under-hand prompts, preference assessment	Researcher	SPED classroom	Case study	Gestures, orienting, verbal, signs, joint attention	Social/turn-taking, requests, commenting, rejecting	Positive effects for multiple behaviors

Thirty-one participants between the ages of 4 and 22 years had significant motor impairments and used wheelchairs for mobility and had severe to profound intellectual disabilities. In all cases, interventions were implemented in special services settings with the exception of one study that took place in an inclusive kindergarten setting (Cosbey & Johnston, 2006). Researchers were the primary communication partners in all 10 studies, except for the study conducted by Cosbey and Johnston in which two young girls with VI learned to interact with peers during center time at school. Communication forms used to activate switches or software included arm, head, or leg movements to activate switches and vocalizations clarified by speech reading software with voice output. All of the communication functions focused on requesting behaviors.

In each of the 10 studies in this category, the adapted devices were tested to ensure that participants could physically activate the switches. Thirty-one students meaningfully used recognizable communication as demonstrated by clear functional relations in six of eight studies that used experimental single-subject designs and positive results for the remainder of the studies in this category. Although 9 of the 10 studies were conducted in specialized settings, Cosbey and Johnston (2006) showed the utility of adapted speech output devices within an inclusive play setting with typical peers, who received minimal coaching from adults as needed, as partners.

Five of the 10 studies were conducted by the same lead researcher, G. E. Lancioni, who used interventions to increase participants' limited repertoire of words or word approximations by using speech recognition software paired with highly preferred matching stimuli. Not only did these interventions rely upon the young adults' production of word approximations or vocalizations, but also they were reinforced by highly preferred stimuli that were recorded messages from friends or family members. Much like clients who participate in speech articulation therapies, these students learned to meaningfully manipulate their voices to improve their speech recognition using a computer software system. Further, participants systematically learned to match their vocal approximations with conceptually matched stimuli, another common type of speech and language activity.

Lancioni and colleagues (Lancioni, Singh, O'Reilly, & Oliva, 2003; Lancioni, Singh, O'Reilly, Oliva, Dardanelli, et al., 2003) also explored participants' abilities to choose specific experiences from a spoken set of preferred options by using their voices or body movements. The emphasis on recognizing verbal messages, using one's own voice to make selections, and the options for participants to select the recorded voices of their grandparents, parents, or friends all contributed to expanding the communication repertoire of these individuals.

3.2. Multicomponent Communication Partner Training Programs

The 12 studies that formed this category were characterized by systematic training of communication partners using a variety of teaching strategies to improve interactions with students. Each study involved an expert who provided specific face-to-face instruction to a communication partner on how to improve interactions with student participants, followed by coaching and feedback. Ten studies involved some elements of video analysis or video-based instruction. In all but one study, child/children's behavioral data from repeated measurement were collected and reported across baseline and intervention conditions. In 11 studies, single-subject design approaches were used, with six using experimental designs. Lappin (2005) used a qualitative case study with the researcher acting as a participant who described qualitative changes in behavior with measurements of frequency for both the baby and his mother.

The ages of participants ranged from infancy to 19 years, although a few additional participants in one study had VI but were older than age 22 (Damen, Kef, Worm, Janssen, & Schuengel, 2011). Thirty-seven students are included within this group. Eight of these 12 studies occurred within a special residential school with residential caregivers and educators. Two occurred in home settings with young children and their mothers (Janssen, Riksen-Walraven, van Dijk, & Ruijssenaars, 2010; Lappin, 2005). One occurred in an inclusive preschool setting with paraeducators and young children (Grisham-Brown, Schuster, Hemmeter, & Collins, 2000). Another was conducted in a special education classroom setting with students and their teachers (Horrocks & Morgan, 2011). While communication forms were described within the studies, the functions of the communication behaviors for both participants and partners were emphasized as a part of the interventions. Such functions included rejecting, requesting, social interaction, informing, and asking. Most of the studies demonstrated positive changes in both partners' and students' communication behaviors, while some participant gains were only seen in specific domains.

One group of studies in the category of multicomponent communication partner training specifically focused on the diagnostic intervention model (DIM) developed by practitioners and researchers in the Netherlands. All nine studies on DIM were led by the same primary researcher,

M. Janssen, except one study in which she had a supporting role (Damen et al., 2011). The goal of DIM is to enhance a communication partner's ability to recognize a child's communication signals, to adjust one's behavior to "attune" or be responsive to the child, and to adapt the interactional context in real time to support the child's connections and participation. The intervention protocol involved five distinct phases that are supported by an analysis of interactions with the students recorded on video. These include team analysis to determine the intervention question (or focus), clarification of the question, interaction analysis (involving eight distinct categories of behavior), implementation of the intervention, and evaluation (Janssen, Riksen-Walraven, van Dijk, Huisman, & Ruijssenaars, 2011). DIM was designed to support students in the context of reviewing their schedules. In this way, the schedule itself, a type of calendar system, could be viewed as an integral part of the communication context. In studies of DIM, the goal was not to teach new communication forms but to increase the frequencies of initiations, exchanges, affective involvement, and shared attention.

In the context of DIM, reductions in challenging behavior were associated with increases in students' ability to regulate interpersonal interactions and intensity. Thus, instead of engaging in self-abusive behavior, turning away from a partner and pausing within the routine, forms of self-regulation, were coded as preferred communication skills. Within the studies on DIM, increased engagement with one another across everyday routines was documented for both caregivers and participants. One exception was noted in the Damen et al. (2011) study in which *t*-tests were used to measure significant changes between baseline and intervention phases. While the intervention produced statistically significant changes for caregivers, students did not increase their initiations.

Other partner training studies included strategies of video-supported instruction, coaching, and feedback to teachers and paraeducators to enable better assessment and intervention within academic settings (Grisham-Brown et al., 2000; Horrocks & Morgan, 2011). Grisham-Brown and colleagues examined an embedded approach to instruction within a preschool setting. Researchers provided specific video training, face-to-face coaching, and training to paraprofessionals in small groups. Four young boys with DB learned to activate switches and use basic sign language to request support or preferred items during center time. They also made receptive gains in following one-step directions during activities of daily living. Much like strategies used in the DIM intervention protocol, video samples of interactions were used with paraeducators to discuss these gains and to examine the paraeducators' roles in supporting these communication behaviors within the environment. In the context of familiar routines and schedule use, paraeducators supported students' receptive and expressive gains.

Using similar instructional strategies, Horrocks and Morgan (2011) implemented a multicomponent partner-based intervention that included seven students with profound multiple disabilities, two who had VI. Teachers received training with video samples, individualized training, step-by-step checklists, and on-the-job coaching to learn both assessment and intervention procedures to use with their students. As a part of the larger intervention that focused on students making gains on a variety of basic academic behaviors, students also were supported to use eye gaze, select objects via touch, and push items away to make basic requests, to say "no," or to attend and engage socially with their teacher. Both students and teachers demonstrated positive gains in response to the intervention.

Finally, in one of the few qualitative intervention approaches, Lappin (2005) taught a therapeutic massage program to a parent who had experienced a negative NICU transition with her baby who was blind. Using naturalistic teaching methods implemented at home, the researcher documented changes in the infant's and mother's interaction from pre- to postintervention using videotaped observations, parent interviews, and a narrative of the process. Prior to intervention, both the mother and the baby demonstrated limited social engagement with each other, and the parent was reported to repeat her negative experiences related to the NICU during most interactions with professionals. Communication forms beyond the therapeutic massage strokes used with her infant included the mother holding her son more frequently, speaking to him more often, and imitating his sounds. The infant, who had been typically still and quiet, increased smiling, laughing, and making happy sounds. Lappin also noted that the baby would draw closer to his mother during and after the intervention phase of this study.

3.3. Tangible Symbols, Object Symbols, and Adapted PECS Interventions

Seven studies, conducted by seven distinct research teams, involved the use of three-dimensional objects or tangible symbols to support students' communication with researchers, peers, or educational staff. Rowland and Schweigert (1989) described object symbols as three-dimensional parts of objects or whole objects that represent things, people, places, actions, or ideas. These symbols were used in the context of children's natural routines and/or with highly preferred activities. Communication partners prompted students with touch, sounds, gestures, or scents to support meaningful exchanges. For all but one of these studies, preference assessment and the use of highly preferred objects or topics were embedded in the instructional design. Three of the studies within this category used pre-/post-designs (Harding et al., 2011; Rowland & Schweigert, 2000; Trief, 2007), and four used experimental single-subject design examinations.

Sixty-one participants with VIMD or DB were included within this group of studies, with 37 of the participants having orthopedic impairments and three children were specifically identified as using wheelchairs. Fifteen of the 61 participants were identified as having autism. Intervention occurred with students in special education settings, with the exception of Rowland and Schweigert (2000), in which 23 students with VI or DB out of a total of 41 received intervention within inclusive and special settings. In all of these studies, intervention happened at naturally occurring times such as snack time or within the class as a preference-based instructional activity.

Researchers primarily served as communication partners except in one study that relied upon teacher reports using checklist data to report implementation and student responses (Trief, 2007). Six of the seven studies emphasized supporting students' expressive communication to make requests of partners using object symbols. Within two of these six studies (Harding et al., 2011; Rowland & Schweigert, 2000), other social functions emerged with partners, including peers. Rowland and Schweigert found that some students did not maintain their skills when the system was used episodically but that for some participants, the strategies were sustained, supporting the development of diverse communication functions with varied partners.

Trief (2007) exposed students to symbols representing typical activities in daily routines. Students showed gains in receptive communication by demonstrating an understanding of object symbols by touching or pointing to one of a pair of object symbols at specific intervals within their routines to answer questions or inform their teachers. Symbol iconicity was found to be a factor for 15 of the 25 students included in the study who consistently demonstrated correct responses.

Four studies within this category adapted the PECS (Bondy & Frost, 1994), which has an established base of evidence for students with autism (Center for Disease Control, 2000). Adaptations were made for students with combined VIs and developmental disabilities, including autism

(Ali et al., 2011; Ivy, Hatton, & Hooper, in press; Lund & Troha, 2008; Parker, Banda, Davidson, & Liu-Gitz, 2010). Changes made to the PECS intervention involved the use of objects or tangible symbols and using sound, smell, or touch to alert students to the presence of reinforcers. Three studies reported generalization of behaviors across communication partners (Ali et al., 2011; Ivy et al., in press) or settings (Parker et al., 2009). For all 12 students, aged 5–17 years, adapted PECS using three-dimensional object symbols was shown to be effective for increasing requesting behaviors.

3.4. Narrative- or Literacy-Based Interventions

Another group of intervention studies focused on building literacy skills. Researchers described changes in communication behaviors using interactive storytelling (Park, 2004) or shared stories and adapted books (Browder, Lee, & Mims, 2011; Mims, Browder, Baker, Lee, & Spooner, 2009). All eight participants were identified with moderate to severe intellectual disabilities. Five of the eight students used wheelchairs. Park did not report student ages; however, the five students attended the Charlton School in the United Kingdom and so were presumed to be younger than 22 years old. Students in the other studies in this category were between 6 and 10 years of age. Two of the three studies employed an experimental single-subject research design (Browder et al., 2011; Mims et al., 2009), and one study employed a qualitative framework to examine participants' engagement in the intervention (Park, 2004).

For each of the studies within this category, communication partners used adapted materials to support a scripted interaction. Communication forms included touching objects, activating voice output switches within a script or narrative framework, and orienting one's body or gaze. Students were supported to answer questions, make statements, or engage in turn-taking while using adapted materials that included props within a dramatic reading, parts of objects within a storybook, and microswitches for narrating parts of a performance. Students made gains in receptive communication skills and expressive behaviors to answer questions or comment. Some students also made gains in social functions of communication such as turn-taking with peers.

3.5. Wait Time

Although several studies described using wait time to support student processing, only Johnson and Parker (2013) studied the effect of wait time on children's responding. Using a single-subject alternating treatment design, Johnson and Parker systematically varied wait time to identify the ideal delay interval for children to increase their responses to verbal prompts in familiar, preferred routines. In this way, the target of the intervention was to increase receptive communication skill. All three students, each with significant disabilities including VI and one with DB, increased responding with wait time. One child within this study was excluded because she was demonstrating greater language fluency and was regularly using close-range American Sign Language, basic writing, and gestures to communicate. The other two participants met criterion as early or emergent communicators. Communication forms included activating toys, pointing to enlarged images of family members, and reaching to identify objects. Communication functions included answering questions and commenting.

3.6. Adapted Prelinguistic Milieu Teaching

A study conducted by Brady and Bashinski (2008) included nine students with DB who were 3–7 years old and attended special education programs. Within this case study, researchers sought to increase participants' repertoire of communicative gestures. Although baseline data were not collected, anecdotal data from parents and educators and researcher observations were used to describe the preintervention communication behaviors of children. Researchers used most-to-least prompting, accessible and reinforcing activities and wait time to support naturalistic conversational exchanges between each child and a researcher during familiar routines. Researchers sought to increase student gestures such as waving hands, nodding head, doing "high fives," shrugging shoulders, and pointing. The frequency and topography of vocalizations, verbalizations, and sign approximations were also measured. Students' communication behaviors were coded from video. The intervention was reported to increase students' frequency of initiations and rates of communication with emphasis on requesting.

4. DISCUSSION

Communication is a dynamic skill, and it remains a highly elusive skill for many students with VIMD. The findings from these 34 intervention studies focusing on the development of emergent communication skills for students with VI and DB demonstrate that varied approaches are being used to support this complex process. From examinations of highly individualized technology (e.g., Lancioni, O'Reilly, et al., 2009; Lancioni, Singh, et al., 2009) to the use of adaptive narratives with scripts (e.g., Browder et al., 2011), these studies offer practical approaches to increase both receptive communication and expressive communication. The purpose of this review is to examine the elements of such interventions to identify implications for professionals and families and directions for future research. Specifically, we examined categories of interventions applied in various contexts with specific communication partners to support students' ability to use various communication forms to address specific communicative functions (Bruce et al., 2004). For discussion, we have synthesized our findings across intervention type to provide a holistic view of the procedures, forms, and functions that have been studied in recent years and the context in which they were studied.

4.1. Student Characteristics and Settings

Although we did not limit our review to students with VIMD, given that our focus was on emergent communication behaviors, perhaps, it is not surprising that almost all of the participants in this set of studies had VIMD. Students with VIMD also represented approximately 65% of young children represented in the Babies Count survey (Hatton et al., 2013) and 90% of the National Deaf-Blind Child Count (Schalock & Bull, 2012). However, given that communication delay is common in children with VIMD, it is disconcerting that researchers are not studying early intervention methods to close the gap between children with and without MD. Additionally, it is unclear why age of students targeted for communication intervention, especially those with multiple disabilities, is relatively old when one considers that (a) earlier intervention is often most effective and (b) infants with severe uncorrectable VI in the United States are being identified at an early age (Hatton et al., 2013). From a developmental perspective, we should be intervening as early as possible.

Though we restricted our review to studies published since 2000, most research was conducted in center-based schools or special education settings. Few studies were conducted in inclusive or home settings. It may be argued that our focus on early and emergent communication behaviors and educational interventions may have influenced this trend, but another explanation may be that conducting intervention research with students who need intensive, supportive instruction to participate in communication in inclusive settings is challenging. Except for a few studies that focused on communication behaviors within the context of literacy instruction, most communication intervention happened within naturalistic routines that included leisure activities and snack times. This trend is consistent with summaries from other reviews that suggest the efficacy of communication intervention is influenced by having highly preferred activities, topics, or treats as a part of the conversation (Parker et al., 2008).

The few studies that implemented intervention in inclusive settings engaged preschool or kindergarten participants in play or self-care activities (Cosbey & Johnston, 2006; Grisham-Brown et al., 2000; Rowland & Schweigert, 2000). For the three studies that took place in home settings, participants were younger than 11 years old. Two such studies involved training participants' mothers to implement intervention (Janssen et al., 2010; Lappin, 2005). The third study was conducted with parents observing the researcher implementing intervention so that family members could adopt strategies found to be successful with their children (Johnson & Parker, 2013). Additionally, two research teams explored intervention efficacy across settings (Janssen et al., 2010; Parker et al., 2009); one of these examined efficacy across communication partners as well (Janssen et al., 2010).

4.2. Intervention Procedures and Outcomes

Communication interventions were designed according to individual capacities, interests, and environmental purposes. Across the diverse interventions being studied, a common theme was the incorporation of assessments and preferences. Before beginning intervention, all of the research teams documented the use of formal and informal assessment tools to assess and describe students' communication prior to beginning intervention. Some studies used elements of standardized assessments, and others included descriptions based on observations or parent reports. Still other studies included video analysis to help teams determine the communication needs of the child within specific contexts (e.g., Janssen et al., 2003). In several studies, formal preference assessments were conducted prior to beginning intervention (e.g., Cosbey & Johnston, 2006; Ivy et al., in press). In studies that used technology to support communication, researchers documented that students could access critical sensory information and make reliable body movements to activate devices (e.g., Lancioni, O'Reilly, et al., 2009; Lancioni, Singh, et al., 2009).

Even though few studies included parents as communication partners, most studies included some aspect of information from parents in the design of the intervention. For example, Johnson and Parker (2013) interviewed parents to gain input on all aspects affecting children's processing speed. Other researchers used recorded messages from parents or grandparents to motivate participants to meaningfully participate. In short, including data from both home and school environments and reviewing assessment data were important components of successful communication interventions. Moreover, the participation of educators and families both as communication partners and as designers of supportive communication systems has been shown to be effective in supporting all students with disabilities (Beukelman & Mirenda, 1998).

Across all studies, several outcomes were targeted, including acquisition of new communicative forms and functions, increased frequency of skills already in children's repertoires, and evolution of existing communicative forms to more sophisticated or recognizable attempts at communication. These outcomes varied across the different types of interventions sometimes involving participants learning a few communication skills or a variety of skills.

4.3. Microswitches, Software, Object Symbols, and PECS

Communication forms such as microswitches, software systems, and object symbols may facilitate easier interpretation of messages and increased communication. Adapting devices or software programs to match the strengths and interests of individuals can reduce isolation and increase quality of life. Microswitch and adapted PECS interventions supported limited communicative functions (e.g., requesting) in circumscribed contexts. Despite the narrow focus of these interventions, students were provided physical access to the environment, predictability and a sense of control, and simple, yet motivating, reasons to communicate. Immediate associations between three-dimensional objects or the spoken words from switch activations and the students' preferred item or activity may have supported participants' memory and access to the communication exchange. Importantly, AAC not only can replace idiosyncratic or undesirable communication forms with more recognizable, socially acceptable communicative forms but also empowers communication partners to respond effectively to students' communicative behaviors, providing shared meaning between students and those around them (Durand, 1993).

For students with VIMD or DB, direct instruction in reaching forward into space to activate a switch or to extend an object cue may serve as a specific support. Movement facilitates the development of communication skills from a conceptual as well as practical standpoint (Bruce, 2005). VI may impede independent, purposeful movement and travel (Lancioni, O'Reilly, van Dijk, & Klaase, 1998).

Additionally, for individuals to be successful communicators, they must learn how to gain a partner's attention to send a message, which often requires a sense of where the potential communication partner is and how to travel to that person to deliver a message (Halle, 1987). Providing incentives for travel to a communication partner in the context of making requests may be a means of increasing students' movement and building a necessary communication skill—locating the communication partner. For these reasons, an adapted form of the PECS (Bondy & Frost, 2001) may be an appropriate communication intervention for some individuals with VIs or DB in that movement, orientation to the partner, and travel are embedded training components that are taught within a highly practical communication context. In summary, the intervention studies included in this review demonstrate that microswitches and object symbols support students' access, interests, and communication.

4.4. Multicomponent Communication Partner Interventions

Multicomponent communication partner interventions emphasize the critical role of communication partners and the conversational "dance" between partner and student. As has been previously described, the ability of partners to recognize, affirm, support, and adapt is a vital component of developing an environment that supports and increases student initiative, engagement, and self-regulation (Vervloed et al., 2006). This group of studies focused primarily on helping adult communication partners be more responsive and used adult learning strategies such as team planning (Janssen, Riksen-Walraven, & van Dijk, 2002), video-based analysis (Janssen et al., 2003), video-based instruction (Horrocks & Morgan, 2011), direct instruction (Grisham-Brown et al., 2000), instruction in natural contexts (Lappin, 2005), onsite coaching (Janssen, Riksen-Walraven, Van Dijk, Huisman, & Ruijssenaars, 2012), incorporating the ideas of adult learners (Janssen et al., 2010), encouragement of adult learners (Janssen et al., 2010; Lappin, 2005), and use of step-by-step materials to support implementation (Horrocks & Morgan, 2011). Beyond partner responsiveness, for communication gains to be sustainable, both the child and the partner may need practical, recognizable forms to use in everyday contexts that support a variety of communication functions.

4.5. Narrative- or Literacy-Based Interventions

In a subset of studies, literacy activities and scripting, in particular, provided students with a framework for participating in communication interactions. Among these studies, predictable patterns (Browder et al., 2011), adapted materials and three-dimensional objects (Mims et al., 2009), and opportunities for turn-taking in a narrative script (Park, 2004) provided participants with contextual support, predictability, reduction of complexity, and concrete referents to support memory.

4.6. Wait Time and Adapted Prelinguistic Milieu Approaches

Wait time is one element often included in larger intervention frameworks. The use of wait time, or processing time, is an important and foundational strategy for supporting communication interactions with students with VIMD and DB (Rowland, 2009). Johnson and Parker (2013) examined it as a solitary intervention element and showed some initial efficacy. From a research standpoint, wait time should be examined in other settings where perhaps there may be more environmental distractions, such as the typical school classroom. Practically, it is one of the intervention elements that may most easily be introduced by parents or teachers in everyday routines.

Brady and Bashinski (2008) used adapted prelinguistic milieu training with nine students with DB over 3 years to increase intentional communication. The intervention was delivered one on one and appeared promising. However, because of the many elements of the intervention—interactions in everyday routines, the use of preferred topics, the intensive support from communication partners, and the use of both objects and touch cues—it is difficult to determine which components may have been most influential in supporting requesting behaviors.

4.7. Limitations

Findings from this review should be considered in light of a few potential limitations of our methods. First, because the scope of this review was limited to early and emergent communication behaviors, some types of communication interventions were not included. For example, studies that focused solely on pragmatics or social skills were not included. Second, studies in which participants had a fluent or recognizable communication system prior to beginning intervention were excluded. Additionally, studies that emphasized intervention solely with communication partners or within the learning environment that did not include child data were also excluded from this review. Third, previous reviews have noted that identifying students with VI or DB in the literature can be challenging because of the many ways in which they may be described (Parker, Davidson, & Banda, 2007). Therefore, we may have overlooked some studies that included members of this diverse population. Finally, it was beyond the scope of this review to rigorously evaluate the quality of research methods.

4.8. Implications for Practitioners and Families

The collective results from the studies reviewed suggest successful and relevant interventions must address individual student needs within specific contexts and with specific partners using accessible methods, messages, and materials. Communication interventions should be expanded to assure accessibility of topics, messages, support materials and tools, and, most importantly, communication partners. Because students with VIMD or DB often struggle to develop contingency awareness and to make connections in their environments, they need support and ongoing encouragement to develop reliable communication systems. For all students with VI who are emergent communicators, preferences should be formally assessed and used to motivate them to communicate.

As other researchers studying AAC have suggested, communication intervention strategies should be based upon unique participant characteristics that can facilitate successful communication in multiple environments (Grunland & Olsson, 1999; Schlosser & Sigafoos, 2002). In this review, researchers tailored the intervention materials to meet the unique needs of the participants. Examples include adapting microswitches to support students' range of motion (e.g., Lancioni et al., 2008), using three-dimensional objects within a story (e.g., Mims et al., 2009), and using parts of highly preferred three-dimensional objects to adapt the PECS intervention (Ivy et al., in press). Furthermore, research findings support that teaching communication skills should be embedded within naturalistic activities throughout each day. For example, Cosbey and Johnston (2006) integrated communication intervention into playtime with peers, part of preschooler's everyday routine.

4.9. Directions for Future Research

Developing early communication skills is a complex process. Compounding this reality with the increasingly fast-paced learning environment and diversity of experiences and needs of children and youth with VI or DB suggests that one type of communication intervention or approach will not work for all individuals, teams, or situations. Therefore, it is important to identify interventions that are resilient and useful for children with diverse learning characteristics and their communication partners in a variety of contexts. For example, children with intellectual disabilities or traumatic brain injury and VI may especially benefit from interventions that maximize and support students' memory. Further, interventions that address the specific needs of students with significant motor disabilities or those with concomitant autism and VI should be identified, with particular attention to how the intervention integrates the participant's strengths, abilities, and interests. Because of the limitless combination of disabilities that may be observed in students with VI, experimental single-subject design is a practical methodology to determine the impact of the intervention (Parker et al., 2007). Experimental single-subject designs also provide the flexibility required to address the sensory, learning, and motivational needs of students within naturalistic settings. Additionally, as many colleagues in AAC research have acknowledged, for the results of an intervention to be sustainable, the communication strategies and supports must be adopted by the communication partners that interact with children every day, which implies that resilient intervention approaches are valued by both partners and the person with emergent communication (Schlosser, Koul, & Costello, 2007). Because the efficacy of an intervention is multidimensional, the research methods should be selected to answer questions about the intensity needed for the intervention to be effective; the practicality of the intervention in everyday environments, with particular attention to how it may be applied with fidelity in naturalistic settings; the maintenance and generalization of the skills learned for both students and communication partners; and, importantly, the social validity of the approach from the family, practitioner, and child's perspectives.

Perhaps the identification of 34 studies over a 13-year period of time is not a modest outcome, when one considers that the goal was to identify interventions for children with a low incidence disability at an emergent phase of communication development. While some of the intervention strategies (e.g., microswitches and communication partner training) were reported in earlier reviews (Parker et al., 2008; Sigafoos et al., 2008), we identified other interventions that are being tested such as intensive teacher training with face-to-face direct instruction, video modeling, checklists, and coaching (Horrocks & Morgan, 2011). We also identified several replications of studies, such as Janssen and colleagues' work using the DIM. Although such replications are encouraging, these interventions need to be replicated by independent researchers in different geographic locations to provide evidence of their efficacy. Within this review, two research teams conducted most of the studies in two categories of intervention, microswitches (Lancioni and colleagues) and multicomponent partner approaches (Janssen and colleagues).

5. CONCLUSION

Within this collection of 34 recent intervention studies for students with VI or DB who are developing emergent communication skills, varied approaches were used to support the complex and dynamic process of promoting communication. Although it was beyond the scope of this examination to rigorously evaluate the quality of research methods, by identifying and organizing the intervention literature into categories, both researchers and practitioners may be able to examine the value and sustainability of approaches for a variety of settings and contexts. Communication itself is one of the most important activities of our lives. Because of its importance and complexity, a variety of research methods are needed to describe, evaluate, and guide the practice of intervention with students who are members of a highly heterogeneous and low incidence population. When the integration of research into practice is balanced by the values of families and students themselves, educators and practitioners have a stronger foundation for implementing evidenced-based communication interventions (ASHA, 2005). The ultimate goal of these communication interventions is to empower students with VI or DB to participate, contribute, and engage with people in the world, not only as an educational goal but also as a fundamental human right.

REFERENCES

Ali, E., MacFarland, S. Z., & Umbreit, J. (2011). Effectiveness of combining tangible symbols with the picture exchange communication system to teach requesting skills to children with multiple disabilities including visual impairment. *Education and Training in Autism and Developmental Disabilities, 46*(3), 425–435.

American Printing House for the Blind (APH). (2007). 2007 annual report. http://74.125.95.132/custom?q=cache:uHvWYUKcI_kJ:www.aph.org/about/ar2007.pdf+annual+report&cd=5&hl=en&ct=clnk&gl=us&client=google-csbeAccessed February 22, 2009.

American Speech-Language-Hearing Association. (2005). *Evidence-based practice in communication disorders [Position Statement].* www.asha.org/policy.

Anderson, E. S., Dunlea, A., & Kekelis, L. S. (1984). Blind children's language: Resolving some differences. *Journal of Child Language, 11*(03), 645–664.

Bates, E., Benigni, L., Bretherton, I., Camaioni, L., & Volterra, V. (1977). *Interaction, conversation, and the development of language.* New York: John Wiley & Sons.

Beukelman, D., & Mirenda, P. (1998). *Augmentative and alternative communication: Management of severe communication disorders in children and adults* (2nd ed.). Baltimore: Brookes.

Bondy, A. S., & Frost, L. A. (1994). The picture exchange communication system. *Focus on Autism and Other Developmental Disabilities*, *9*(3), 1–19.
Bondy, A., & Frost, L. (2001). The Picture Exchange Communication System. *Behavior Modification*, *25*(5), 725–744.
Brady, N. C., & Bashinski, S. M. (2008). Increasing communication in children with concurrent vision and hearing loss. *Research and Practice for Persons with Severe Disabilities: The Journal of TASH*, *33*(1–2), 59.
Broadston, P. M. (2003). *The functional hearing inventory: Criterion-related validity interrater validity*. Doctoral dissertation, Texas Tech University.
Browder, D. M., Lee, L., & Mims, P. (2011). Using shared stories and individual response modes to promote comprehension and engagement in literacy for students with multiple, severe disabilities. *Education and Training in Autism and Developmental Disabilities*, *46*(3), 339–351.
Bruce, S. M. (2005). The application of Werner and Kaplan's concept of "distancing" to children who are deaf-blind. *Journal of Visual Impairment & Blindness*, *99*, 1–27.
Bruce, S., Godbold, E., & Naponelli-Gold, S. (2004). An analysis of communicative functions of teachers and their students who are congenitally deafblind. *RE:view: Rehabilitation Education for Blindness and Visual Impairment*, *36*(2), 81.
Center for Disease Control. (2000). *Prevalence of autism spectrum disorders: Autism and Developmental Disabilities Monitoring Network, 14 Sites, United States, 2002*. Atlanta, GA: Centers for Disease Control and Prevention.
Chen, D., Klein, M., & Haney, M. (2000). *Promoting learning through active interaction: A guide to early communication with young children who have multiple disabilities*. Baltimore: Paul H. Brookes.
Cosbey, J. E., & Johnston, S. (2006). Using a single-switch voice output communication aid to increase social access for children with severe disabilities in inclusive classrooms. *Research and Practice for Persons with Severe Disabilities*, *31*(2), 144–156.
Damen, S., Kef, S., Worm, M., Janssen, M. J., & Schuengel, C. (2011). Effects of video-feedback interaction training for professional caregivers of children and adults with visual and intellectual disabilities. *Journal of Intellectual Disability Research*, *55*(6), 581–595.
Dote-Kwan, J., & Hughes, M. (1994). The home environment of young blind children. *Journal of Visual Impairments & Blindness*, *88*, 31–43.
Durand, V. M. (1993). Functional communication training using assistive devices: Effects on challenging behavior and affect. *Augmentative and Alternative Communication*, *9*, 168–176.
Ferrell, K. A. (2000). Growth and development of young children with visual impairments. In M. C. Holbrook, & A. J. Koenig (Eds.), *Foundations of education: Volume I. History and theory of teaching children and youths with visual impairments* (pp. 111–134) (2nd ed.). New York: AFB Press.
Grisham-Brown, J., Schuster, J. W., Hemmeter, M. L., & Collins, B. C. (2000). Using an embedding strategy to teach preschoolers with significant disabilities. *Journal of Behavioral Education*, *10*(2/3), 139–162.
Grove, N., Bunning, K., Porter, J., & Olsson, C. (1999). See what I mean: Interpreting the meaning of communication by people with severe and profound intellectual disabilities. *Journal of Applied Research in Intellectual Disabilities*, *12*(3), 190–203.
Grunland, M., & Olsson, C. (1999). Efficacy of communication intervention for presymbolic communicators. *Augmentative and Alternative Communication*, *15*(1), 25–37.
Halle, J. W. (1987). Teaching language in the natural environment: an analysis of spontaneity. *JASH*, *12*(1), 28–37.
Harding, C., Lindsay, G., O'Brien, A., Dipper, L., & Wright, J. (2011). Implementing AAC with children with profound and multiple learning disabilities: A study in rationale underpinning intervention. *Journal of Research in Special Educational Needs*, *11*(2), 120–129.

Hatton, D. D., Ivy, S. E., & Boyer, C. (2013). Severe visual impairments in infants and toddlers in the United States. *Journal of Visual Impairment & Blindness, 107*(5), 325–336.
Heller, K. W., & Allgood, M. H. (1996). Use of dual communication boards at vocational sites by students who are deaf-blind. *RE:view, 27*, 180–191.
Horner, R. H., Carr, E. G., Halle, J., Odom, S., & Wolery, M. (2005). The use of single subject research to identify evidence-based practice in special education. *Exceptional Children, 71*, 165–179.
Horrocks, E. L., & Morgan, R. L. (2011). Effects of inservice teacher training on correct implementation of assessment and instructional procedures for teachers of students with profound multiple disabilities. *Teacher Education and Special Education, 34*(4), 283–319.
Hourcade, J., Pilotte, T. E., West, E., & Parette, P. (2004). A history of augmentative and alternative communication for individuals with severe and profound disabilities. *Focus on Autism and other Developmental Disabilities, 19*(4), 235–244.
House, S. S., & Davidson, R. C. (2000). Speech-language pathologists and children with sensory impairments. *Communication Disorders Quarterly, 21*(4), 224.
Ivy, S. E., Hatton, D. D., & Hooper, J. D. (in press). Using the Picture Exchange Communication System with students with visual impairment. *Exceptional Children.*
Janssen, M. J., Riksen-Walraven, J. M., & van Dijk, J. P. (2002). Enhancing the quality of interaction between deafblind children and their educators. *Journal of Developmental and Physical Disabilities, 14*(1), 87–109.
Janssen, M. J., Riksen-Walraven, J. M., & van Dijk, J. P. (2003). Contact: Effects of an intervention program to foster harmonious interactions between deaf-blind children and their educators. *Journal of Visual Impairment & Blindness, 97*(4), 215–229.
Janssen, M. J., Riksen-Walraven, J. M., & van Dijk, J. P. (2004). Enhancing the interactive competence of deafblind children: Do intervention effects endure? *Journal of Developmental and Physical Disabilities, 16*(1), 73–94.
Janssen, M., Riksen-Walraven, J. M., & van Dijk, J. P. (2006). Applying the diagnostic intervention model for fostering harmonious interactions between deaf-blind children and their educators: A case study. *Journal of Visual Impairment & Blindness, 100*(2), 91–105.
Janssen, M. J., Riksen-Walraven, J. M., van Dijk, J. P., Huisman, M., & Ruijssenaars, W. A. (2011). Fostering harmonious interactions in a boy with congenital deafblindness: A single-case study. *Journal of Visual Impairment & Blindness, 105*(9), 560–572.
Janssen, M. J., Riksen-Walraven, J. M., Van Dijk, J. P., Huisman, M., & Ruijssenaars, W. A. (2012). Enhancing sustained interaction between children with congenital deafblindness and their educators. *Journal of Visual Impairment & Blindness, 106*(3), 177–183.
Janssen, M. J., Riksen-Walraven, J. M., van Dijk, J. P., & Ruijssenaars, W. A. (2010). Interaction coaching with mothers of children with congenital deaf-blindness at home: Applying the diagnostic intervention model. *Journal of Visual Impairment & Blindness, 104*(1), 15–29.
Janssen, M. J., Riksen-Walraven, J. M., Van Dijk, J. P., Ruijssenaars, W. A., & Vlaskamp, C. (2007). Team interaction coaching with educators of adolescents who are deaf-blind: Applying the diagnostic intervention model. *Journal of Visual Impairment & Blindness, 101*(10), 677–689.
Johnson, N., & Parker, A. (2013). Effects of wait time when communicating with children who have sensory and additional impairments. *Journal of Visual Impairment & Blindness, 107*(5), 363–374.
Johnston, S. S., Reichle, J., & Evans, J. (2004). Supporting augmentative and alternative communication use by beginning communicators with severe disabilities. *American Journal of Speech-Language Pathology, 13*, 20–30.
Lancioni, G. E., O'Reilly, M. F., van Dijk, J., & Klaase, M. (1998). An environmental enrichment programme to promote adaptive responding in two children with multiple disabilities. *Scandinavian Journal of Behaviour Therapy, 27*(3), 130–134.

Lancioni, G. E., O'Reilly, M. F., Singh, N. N., Sigafoos, J., Didden, R., Oliva, D., et al. (2009). Persons with multiple disabilities accessing stimulation and requesting social contact via microswitch and VOCA devices: New research evaluation and social validation. *Research in Developmental Disabilities, 30*(5), 1084–1094.
Lancioni, G. E., O'Reilly, M. F., Singh, N. N., Sigafoos, J., Oliva, D., Montironi, G., et al. (2005). Extending the evaluation of a computer system used as a microswitch for word utterances of persons with multiple disabilities. *Journal of Intellectual Disability Research, 49*(9), 639–646.
Lancioni, G. E., O'Reilly, M. F., Singh, N. N., Sigafoos, J., Oliva, D., & Severini, L. (2008). Three persons with multiple disabilities accessing environmental stimuli and asking for social contact through microswitch and VOCA technology. *Journal of Intellectual Disability Research, 52*(4), 327–336.
Lancioni, G. E., Singh, N. N., O Reilly, M. F., Oliva, D., & Montironi, G. (2004). A computer system serving as a microswitch for vocal utterances of persons with multiple disabilities: Two case evaluations. *Journal of Visual Impairment & Blindness, 98*(2), 116–120.
Lancioni, G. E., Singh, N. N., O'Reilly, M. F., & Oliva, D. (2003). Extending microswitch-based programs for people with multiple disabilities: Use of words and choice opportunities. *Research in Developmental Disabilities, 24*(2), 139–148.
Lancioni, G. E., Singh, N. N., O'Reilly, M. F., Oliva, D., Dardanelli, E., & Pirani, P. (2003). Adapting the use of microswitches to foster response awareness and word association: Two case evaluations. *Journal of Positive Behavior Interventions, 5*(3), 153–157.
Lancioni, G. E., Singh, N. N., O'Reilly, M. F., Sigafoos, J., Oliva, D., & Montironi, G. (2004). Evaluating a computer system used as a microswitch for word utterances of persons with multiple disabilities. *Disability and Rehabilitation: An International, Multidisciplinary Journal, 26*(21/22), 1286–1290.
Lancioni, G. E., Singh, N. N., O'Reilly, M. F., Sigafoos, J., Oliva, D., & Baccani, S. (2006). Teaching 'yes' and 'no' responses to children with multiple disabilities through a program including microswitches linked to a vocal output device. *Perceptual and Motor Skills, 102*(1), 51–61.
Lancioni, G. E., Singh, N. N., O'Reilly, M. F., Sigafoos, J., Oliva, D., & Cingolani, E. (2009). Students with multiple disabilities using technology-based programs to choose and access stimulus events alone or with caregiver participation. *Research in Developmental Disabilities, 30*(4), 689–701.
Lappin, G. (2005). Using infant massage following a mother's unfavorable neonatal intensive care unit experiences: A case study. *RE:view: Rehabilitation Education for Blindness & Visual Impairment, 37*(2), 87–94.
Lund, S. K., & Troha, G. M. (2008). Teaching young people who are blind and have autism to make requests using a variation on the picture exchange communication system with tactile symbols: A preliminary investigation. *Journal of Autism Development Disorder, 38*, 719–730.
Mason, C., & Davidson, R. (2001). *National plan for training personnel to serve children with blindness and low vision.* Reston, VA: Council for Exceptional Children.
Mathy-Laikko, P., Iacono, T., Ratcliff, A., Villarruel, F., Yoder, D., & Vanderheiden, G. (1989). Teaching a child with multiple disabilities to use a tactile augmentative communication device. *Augmentative and Alternative Communication, 5*, 249–256.
Millar, D. C., Light, J. C., & Schlosser, R. W. (2006). The impact of augmentative and alternative communication intervention on the speech production of individuals with developmental disabilities: A research review. *Journal of Speech, Language, and Hearing Research, 49*, 248–264.
Mims, P. J., Browder, D. M., Baker, J. N., Lee, A., & Spooner, F. (2009). Increasing comprehension of students with significant intellectual disabilities and visual impairments during shared stories. *Education and Training in Developmental Disabilities, 44*(3), 409.

Odom, S. L., Brantlinger, E., Gersten, R., Horner, R. D., Thompson, B., & Harris, K. (2004). *Quality indicators for research in special education and guidelines for evidence-based practices: Executive summary*. Arlington, VA: Council for Exceptional Children Division for Research.

Odom, S. L., Brantlinger, E., Gersten, R., Horner, R. H., Thompson, B., & Harris, K. R. (2005). Research in special education: Scientific methods and evidence-based practices. *Exceptional Children, 71*, 137–148.

Park, K. (2004). Interactive storytelling: From the book of genesis. *British Journal of Special Education, 31*(1), 16–23.

Parker, A. T., Davidson, R. C., & Banda, D. R. (2007). Emerging evidence from single subject research in the field of deaf-blindness. *Journal of Visual Impairment and Blindness, 101*, 690–700.

Parker, A. T., Devender, B. R., Davidson, R. C., & Liu-Gitz, L. (2009). Adapting the picture exchange communication system for a student with visual impairment and autism: A case study. *Research and Practice in Visual Impairment & Blindness, 3*(1), 2–11.

Parker, A. T., Grimmett, E., & Summers, S. (2008). Evidence-based communication practices for children with vision loss and additional disabilities: An examination of single-subject design studies. *Journal of Visual Impairment & Blindness, 102*, 540–552.

Parker, A. T., Banda, D. R., Davidson, R., & Liu-Gitz, L. (2010). Adapting the Picture Exchange Communication System (PECS) for a student with visual impairment and autism: A case study. *AER Journal: Research and Practice in Visual Impairment and Blindness, 3*(1), 2–11. Winter, 2010.

Pérez-Pereira, M., & Castro, J. (1997). Language acquisition and the compensation of visual deficit: New comparative data on a controversial topic. *British Journal of Developmental Psychology, 15*(4), 439–459.

Ronnberg, J., & Borg, E. (2001). A review and evaluation of research on the deaf-blind from perceptual, communicative, social and rehabilitative perspectives. *Scandinavian Audiology, 30*, 67–77.

Rowland, C. (1990). Communication in the classroom for children with dual sensory impairments: Studies of teacher and child behavior. *Augmentative and Alternative Communication, 6*(4), 262–274.

Rowland, C. (2009). *Assessing communication and learning in young children who are deafblind or who have multiple disabilities*. Portland, OR: Oregon Institute on Disability and Development.

Rowland, C., & Schweigert, P. (1989). Tangible symbols: Symbolic communication for individuals with multisensory impairments. *Augmentative and Alternative Communication, 5*(4), 226–234.

Rowland, C., & Schweigert, P. (2000). Tangible symbols, tangible outcomes. *Augmentative and Alternative Communication, 16*(2), 61–78.

Rowland, C., & Stremel-Campbell. (1987). Share and share alike: Conventional gestures to emergent language for learners with sensory impairments. In L. Goetz, D. Guess, & K. Stremel-Campbell (Eds.), *Innovative program design for individuals with dual sensory impairments* (pp. 49–75). Baltimore: Paul H. Brookes Publishing Company.

Schalock, M. D., & Bull, R. (2012). *The 2011 National Child Count of children and youth who are deaf-blind*. Monmouth, OR: National Consortium on Deaf-blindness, Teaching Research Institute, Western Oregon University.

Schlosser, R. W., Koul, R., & Costello, J. (2007). Asking well-built questions for evidence-based practice in augmentative and alternative communication. *Journal of Communication Disorders, 40*(3), 225–238.

Schlosser, R. W., & Sigafoos, J. (2002). Selecting graphic symbols for an initial request lexicon: Integrative review. *Augmentative and Alternative Communication, 18*, 102–123.

Schweigert, P. (1989). Use of microswitch technology to facilitate social contingency awareness as a basis for early communication skills. *Augmentative and Alternative Communication, 5*(3), 192–198.

Seligman, M. (1975). *Helplessness: On depression, development and death.* San Francisco: W.H. Freeman.

Siegel-Causey, E., & Downing, J. (1987). Nonsymbolic communication development: Theoretical concepts and educational strategies. In L. Goetz, D. Guess, & K. Stremel-Campbell (Eds.), *Innovative program design for individuals with dual sensory impairments* (pp. 191–223). Baltimore: Paul H. Brookes.

Sigafoos, J. (2005). From Premack to PECS: 25 years of progress in communication intervention for individuals with developmental disabilities. *Educational Psychology, 25*(6), 601–607.

Sigafoos, J., Didden, R., Schlosser, R., Green, V. A., O'Reilly, M. F., & Lancioni, G. E. (2008). A review of intervention studies on teaching AAC to individuals who are deaf and blind. *Journal of Developmental and Physical Disabilities, 20*, 70–99.

Singh, N. N., Lancioni, G. E., O'Reilly, M. F., Molina, E. J., Adkins, A. D., & Oliva, D. (2003). Self-determination during mealtimes through microswitch choice-making by an individual with complex multiple disabilities and profound mental retardation. *Journal of Positive Behavior Interventions, 5*, 209–215.

Trief, E. (2007). The use of tangible cues for children with multiple disabilities and visual impairment. *Journal of Visual Impairment & Blindness, 101*(10), 613–619.

Turnell, R., & Carter, M. (1994). Establishing a repertoire of requesting for a student with severe and multiple disabilities using tangible symbols and naturalistic time delay. *Journal of Intellectual and Developmental Disability, 19*(3), 193–207.

U.S. Department of Education: Office of Special Education and Rehabilitative Services. (2001). *Information on the provision of vocational rehabilitation services to individuals with significant speech and language impairments.* Washington, DC: US. Rehabilitation Services Administration.

van Dijk, J. (1969). *Educational approaches to abnormal development: Some aspects of the psychology of the senses in relation to the sensory-deprived child.* Dallas, TX: South Central Regional Center for Services to the Deaf-Blind, 2930 Turtle Creek Plaza.

Vervloed, M., van Dijk, R., Knoors, H., & van Dijk, J. (2006). Interaction between the teacher and the congenitally deafblind child. *American Annals of the Deaf, 151*(3), 336–344.

Warburg, M. (2008). Visual impairment in adult people with intellectual disability: Literature review. *Journal of Intellectual Disability Research, 45*(5), 424–438.

Warren, D. H. (1994). *Blindness and children: An individual differences approach.* New York, NY: Cambridge University Press.

Warren, D. H., & Hatton, D. D. (2003). Cognitive development of children with visual impairments. In I. Rapin & S. Segalowitz Eds., *Part II: Child neuropsychology Vol. 8. Handbook of neuropsychology* (pp. 439–458) (2nd ed.). New York, NY: Elsevier.

Werner, H., & Kaplan, B. (1964). *Symbol formation: An organismic-developmental approach to language and expression of thought.* London: John Wiley & Sons.

CHAPTER FIVE

Unified English Braille in the United States: A Research Agenda for Transition and Instruction

Frances Mary D'Andrea[*,1], **Diane P. Wormsley**[†], **Mackenzie E. Savaiano**[‡]
[*]Braille Authority of North America, Pittsburgh, Pennsylvania, United States
[†]Department of Curriculum and Instruction, North Carolina Central University, Durham, North Carolina, United States
[‡]Department of Special Education, Vanderbilt University, Nashville, Tennessee, USA
[1]Corresponding author: e-mail address: literacy2@mindspring.com

Contents

Abstract

The authors examined research on Unified English Braille (UEB) in the United States and other countries to provide researchers in the field of blindness and visual impairment with a research agenda for the transition to UEB in the United States. The search revealed 25 research articles, peer-reviewed and non-peer-reviewed, on UEB and 7 reports on UEB implementation in other countries. Overall, the research covered (a) physical attributes of UEB, (b) perceptions of UEB, and (c) readability of UEB. Although studies were few and varied and direct comparisons difficult, valuable information was revealed about implications for code change in the United States. A great deal of the literature

International Review of Research in Developmental Disabilities, Volume 46
ISSN 2211-6095
http://dx.doi.org/10.1016/B978-0-12-420039-5.00007-1

examined opinions and attitudes toward the adoption of UEB and suggested that implementation in most countries was accomplished sooner than projected and that the transition was not as disruptive as anticipated. Although the United States can use resources, research, and experiences from other countries that have already made the transition to UEB, we should consider differences between the United States and other countries, including the implications of being the only International Council on English Braille member country to maintain a separate code for mathematics.

1. INTRODUCTION

In November 2012, the Braille Authority of North America (BANA) voted to adopt the Unified English Braille (UEB) as a code for use in the United States. BANA is currently in the midst of developing an implementation plan and transition time line to make the change from the current code, *English Braille American Edition* (EBAE), to UEB.

This chapter will examine the implications of this decision. First, a brief history of braille use in the United States will provide some background to understand the context in which the decision was made. Next, the authors will provide a review of the available research and other published papers regarding UEB, in both the United States and internationally. Results of this literature review will provide a framework for the discussion of the development of a research agenda in the United States regarding UEB.

1.1. Tactile Reading Through the Years

Prior to the late 1700s, children who were blind generally did not attend school; in some instances, wealthy families provided tutors who invented methods for reading tactually, such as the use of pinpricks or letters carved into wood (Catholic Encyclopedia, n.d.). The first school for the blind was opened in 1784 in Paris, France, by Valentin Haüy, *L'Institut National des Jeunes Aveugles*. The school was so successful that it was swiftly followed by schools for the blind being established throughout Europe, with schools being founded in England in 1791, Scotland in 1793, Austria–Hungary in 1804, Germany in 1806, Russia in 1807, Sweden in 1808, Switzerland in 1809, and Spain in 1820. The United States followed suit in 1831 and Canada in 1861 (D'Andrea, 2009).

Generally, these schools taught their students to read embossed print, although the limitations of that method were widely known. Embossed print books were bulky and hard to read and difficult and expensive to create. In addition, embossed print was not a system that the blind students themselves could write independently or easily. Educators in many countries sought better solutions for tactile reading and writing. Most of the systems developed at that time were based on print letters, as some educators felt that any system that did not involve tactile print would set blind people apart from society (Irwin, 1955). For example, in 1831, James Gall of Edinburgh, Scotland, invented an angular Roman type of embossed print to publish a volume of the Gospel of St. John. Other systems were devised that combined elements of simplified print. Probably, the system that lasted the longest was Moon type, developed by William Moon in 1847. This simplified system of tactile shapes, many loosely based on print letters, was used in the United Kingdom and in the United States. It is still used to a very small extent today, primarily in England, and supported by an independent group (Royal National Institute of Blind People, 2012).

The braille code was invented in 1829 by Louis Braille, a student at *L'Institut National des Jeunes Aveugles*; Braille published revisions to his original code in 1834 and 1837 (Mellor, 2006). Unlike other tactile codes of the time, Braille's system was based on a "cell" of six dots arrayed in two columns of three dots each. Braille reportedly got the idea to use dots after seeing a system of "night writing" developed by Charles Barbier and used in the French military. Barbier's system was based on phonetics rather than the alphabet and used 12 dots instead of six. Braille's improvements allowed tactile readers to use standard spelling and to read and write books quickly and easily. The smaller cell allowed the reader to feel the entire letter under his or her fingertips. Braille also invented a tactile system for music notation based on the 6-dot cell. There was some initial reluctance by school officials to adopt Braille's system since it did not look like print and thus was not readily readable by print users; however, the system was enthusiastically adopted by the students at *L'Institut*. The use of the system spread quickly even after Braille's untimely death in 1847. The braille code was adopted in France by 1854, and its use had extended to other countries by the end of the 1800s (Lorimer, 2000).

1.2. Tactile Reading in the United States

The first book to be embossed the United States was published in 1833 at the Pennsylvania Institution for the Instruction of the Blind, using embossed print (referred to as a "line type"). A similar system was called Boston line type, which was used to publish a book in 1834. Boston line type went on to become quite popular in the United States for the next 50 years (Bogart, 2000).

The braille code was first used in the United States in 1866, when Dr. Simon Pollock, superintendent of the Missouri School for the Blind, introduced the system after traveling in Europe. However, he found competition for the use of braille from other systems such as Boston line type, New York Point, and American Braille. New York Point, developed by William Bell Wait, superintendent of the New York Institution for the Education of the Blind, was a dot system presented horizontally and based on letter frequency; more frequently used letters were "shorter" than less common letters and so saved space on the page. Another competing system developed in 1878 by J. W. Smith called the American Modified Braille was also based on letter frequency but used the 3 × 2-dot cell (Irwin, 1955).

By the end of the 1800s, materials were being produced in a number of tactile systems that competed for prominence, championed by educators from different schools for students who were blind and by students and readers who had learned specific codes. Proponents of the various codes were steadfast in their opinion that their favorite code was best for readers who were blind. This competition resulted in wasted and duplicative effort as popular books had to be embossed in multiple systems. People who were blind often had to learn more than one tactile code to read books and magazines produced from various printing houses in the United States and imported from England (which had used braille since the 1870s). There was widespread recognition that adopting a standard tactile code would be advantageous for readers and producers of materials.

Several "uniform type" committees were established in the early years of the twentieth century to settle this "war of the dots," but it was not until 1918 that this contentious issue was settled—for the time being (Irwin, 1955). The United States adopted the Revised Braille (also referred to as "grade 1-1/2") that used the same alphabet used in countries using Braille's system, but with only 44 contractions as opposed to 189 contractions used in England. Children in the United States were initially taught the braille alphabet, and words were spelled out until about third grade when the few contractions were introduced (Maxfield, 1928). However, in 1929, only 12 years after seeming to settle the issue about which code to use, educators were already questioning whether to add more contractions to the code used in the United States to match the system used in the United Kingdom. A joint committee of delegates from the United States and Britain met in 1932 to resolve most of the differences between the codes used in the two countries. The "Treaty of London" resulted in the adoption of Standard English Braille. However, some differences in contractions, capitalization, and rules for contractions persisted. The literary code was not uniform internationally (Bogart, 2000).

Over the decades following the Treaty of London, the literary code, EBAE, continued to grow and be revised. As braille reflects a living language, it must change periodically to reflect new standards of print. The current version of EBAE, while it has a copyright of 1994, had major revisions in 2002 and most recently in 2007. The current version of EBAE is the basis of guidelines for formats, tactile graphics, and foreign language transcription.

1.3. Systems for Mathematics and Scientific Notation

So far, this chapter has examined the history of what is referred to as "literary braille," but there was wide understanding that a tactile system for mathematics was also needed. Unfortunately, relatively little can be found in the literature about the development of tactile codes for mathematics compared to tactile codes for literary materials.

Illingsworth (1910) barely mentioned arithmetic in his book *History of the Education of the Blind* but described a tangible aid called the "Taylor's Octagonal Board and Pegs" (based on work by Rev. William Taylor) that was widely used for students to perform mathematical problems from basic operations to algebra. Later, a blind mathematician at Cambridge University named Henry Martyn Taylor developed a system of mathematics notation based on the 6-dot cell (New Beacon, 1934). The Taylor code was used for mathematics in Britain and the United States although the limitations of that system led another international committee to consider various codes (such as the German Marburg code) to replace it. In the early 1950s, the Joint Uniform Braille Committee's subcommittee on mathematics considered the merits of revising the Taylor code again, adopting a revised international code (as used in Britain) or the newly created Nemeth code. In November 1954, the American Printing House adopted the Nemeth code for mathematics transcription (Hooper, 1957). The code was modified in 1965 and once again in 1972 (Braille Authority of North America, 1972). The overall process of adopting and implementing the Nemeth code in the United States is not recorded in the literature, although a survey of residential schools conducted in 1970 regarding mathematics instruction noted that one school reported the need for simplification of the Nemeth code and for the code to become standard without further changes (Lewis, 1970). The code has remained stable and substantially unchanged since the 1972 version although, like any code used in the United States, periodic updates, clarifications, and adjustments have been made. A chemistry code based on the Nemeth code was adopted in 1997.

Yet another technical code, the Computer Braille Code (CBC), was adopted by BANA in 1987 to support the burgeoning use of computers to create electronic texts that could be translated into braille. These texts could then be embossed onto paper or could be read as "paperless" braille (Grey, 2009). The development of CBC introduced an additional character set for users.

1.4. Search for a Unified Code

By the late 1980s, there was increasing concern that the use of four different braille codes in the United States was unwieldy and overly complex. The surfeit of braille codes meant that common print symbols such as the dollar sign were rendered differently depending on context—one dollar sign for "literary" braille materials, another for mathematics textbooks, and yet another for computer notation. The line between what was considered literary and technical material was blurring, with URLs and e-mail addresses commonly occurring in general interest material and primary textbooks; no longer was "computer notation" a code only for highly skilled adults. The transcription of braille was complicated because accurate translation and backtranslation (from braille back to print) was difficult to achieve. As the English language changed, the braille code did not have the capacity to keep up with changes in new usage (e.g., periods used in the middle of words or names such as popular musician will.i.am) and in developing new symbols (Braille Authority of North America, n.d.).

In 1991, Dr. Tim Cranmer and Dr. Abraham Nemeth addressed the BANA Board with a letter outlining their concerns about braille codes in North America and proposed a project to develop a unified code, one that would encompass symbols and rules to allow for literary and technical transcription (Bogart, Cranmer, & Sullivan, 2000). The BANA Board voted to start a research project to see if such a code was feasible. Soon after, the project caught the interest of the International Council on English Braille (ICEB) and work on the Unified Braille Code (soon after renamed Unified English Braille, UEB) became a major international project. Members of the various UEB committees were predominantly braille readers and work was conducted primarily via the Internet, although a few face-to-face meetings were also held.

One of the most important early decisions that the committee had to make was to identify the mathematics system that would be used in the unified code. The committee considered three different systems: upper numbers (numbers formed in the upper part of the braille cell, as used in literary contexts), lower numbers (numbers formed in the lower part of the cell, such as those used in the Nemeth code), and Antoine numbers (numbers formed with a dot-6, as used in France). After a lengthy deliberation, the committee decided to use upper numbers due to their overall familiarity to readers internationally, their ease of reading, and a preliminary analysis indicating that number/punctuation combinations were more frequent than number/letter combinations. Upper numbers were considered less disruptive to the code since they would not interfere with punctuation marks, which are in the lower part of the cell (Bogart et al., 2000).

By the 2004 ICEB General Assembly, the code was determined to be sufficiently complete for each member country to vote on its adoption. Over the past decade, all the ICEB countries have adopted UEB: first, South Africa and Australia and then New Zealand, Nigeria, Canada, and the United Kingdom. Although the project originated in the United States, it was the last to adopt UEB, voting for adoption in November 2012. However, unlike the other ICEB member countries, the vote in the United States contained the caveat that the Nemeth code would be maintained. It is also interesting to note that Canada, a BANA member, and New Zealand, an associate member of BANA, voted to adopt UEB well before the United States. New Zealand's implementation is essentially complete, and very few people are still using the older codes replaced by UEB (i.e., EBAE and the Nemeth code). While to date Canada is still developing its implementation plan, it does not plan to continue using the Nemeth code.

Prior to the implementation of UEB in the United States, the authors decided to examine the research on UEB in other countries and in the United States and implementation reports from countries that had already adopted UEB. The purpose of this chapter is to suggest a research agenda for the transition to UEB in the United States based on current literature on UEB.

2. METHOD

We included in this chapter articles published between 1998 and 2013 that reported research on UEB and were written in English and articles or reports that discussed the evaluation or implementation of UEB. A search was conducted using ERIC, PsycINFO, ProQuest Central, and ProQuest

Dissertations and Theses databases using the terms "unified" and "braille," and a substantial number of citations (323) were identified by the search. Upon closer inspection, 156 (48%) of the citations returned from the databases were neuroscience articles. Braille reading is a widely used example of the brain's ability to reorganize and make new connections (Kamei-Hannan, 2006), but these were not articles about braille reading. Additionally, 20 (6%) citations were from library science or information technology journals that used braille as an example of a necessary accommodation.

After applying inclusion criteria to the literature identified by the search, 14 articles were about UEB. Of those articles, seven were eliminated because they were opinions or discussions of theories, beliefs, or practices. As articles were identified as meeting inclusion criteria, their reference lists were checked to identify further articles of interest. In addition, searches of the following periodicals were conducted by hand to ensure inclusion of all relevant articles: *AFB Research Bulletin, British Journal of Visual Impairment, Education of the Visually Handicapped, International Journal of Education for the Blind, Journal of Visual Impairment & Blindness, New Outlook for the Blind, and RE:view*. Two additional articles were identified from published materials. Nine articles returned by these methods met all inclusion criteria.

An additional hand search was conducted of the websites of the following professional organizations and associations related to braille use internationally: *Anglo-Nigerian Welfare Association for the Blind, Australian Braille Authority, Braille Authority of New Zealand Aotearoa Trust, Braille Authority of North America, Braille Authority of the United Kingdom, Canadian Braille Authority, Canadian National Institute for the Blind, Royal National Institute of Blind People, International Council on English Braille*, and *South African National Council for the Blind*. Lastly, the proceedings of the most recent meeting of the ICEB were reviewed for information on implementation. Using these methods, 15 additional research reports were identified and 7 implementation reports from the ICEB General Assembly. A total of 25 research articles (see Table 5.1) and 7 implementation reports were included in the review in the succeeding text.

3. RESULTS

The results of our literature search presented us with a number of challenges. First, the search revealed a paucity of literature on UEB. Grouping what we did find resulted in peer-reviewed articles and non-peer-reviewed articles or reports. The largest portion of the non-peer-reviewed materials

Table 5.1 Summary of Studies Included in Review

Study	Country	Category	# Participants/ Respondents	Design/ Methodology
Peer-Reviewed				
Bogart and Koenig (2005)	Australia, Canada, Japan, New Zealand, Nigeria, South Africa, the United Kingdom, the United States	Perceptions	446	Quantitative: questionnaire
Cryer, Home, and Morley Wilkins (2013a)	The United Kingdom	Readability	6 Technical experts	Qualitative: observations and interviews
Cryer, Home, and Morley Wilkins (2013b)	The United Kingdom	Perceptions	27 Braille readers, transcribers, and teachers (21 in focus groups)	Qualitative: focus groups
Gerber and Smith (2006)	Canada	Perceptions	85 (49 teachers, 36 students)	Qualitative: focus groups
Holbrook and MacCuspie (2010)	Canada and the United States	Readability	5	Qualitative: observations and focus groups
Knowlton and Wetzel (2006)	The United States	Physical attributes	NA (analysis of literary/ technical materials)	Quantitative: descriptive
Steinman, Kimbrough, Johnson, and LeJeune (2004)	The United States	Readability	8	Quantitative: pretest/ posttest
Wetzel and Knowlton (2006a)	The United States	Perceptions	79	Qualitative: questionnaire and focus groups
Wetzel and Knowlton (2006b)	The United States	Readability	52	Quantitative: 3 × 2 × 4 ANOVA

Continued

Table 5.1 Summary of Studies Included in Review—cont'd

Study	Country	Category	# Participants/ Respondents	Design/ Methodology
Non-Peer-Reviewed				
Bogart, D'Andrea, and Koenig (2004)	Canada	Physical attributes	NA (analysis of literary/ technical materials)	Quantitative: descriptive
Braille Authority of North America (2003)	The United States and Canada	Perceptions	248	Quantitative: questionnaire
Braille Authority of the United Kingdom (2008)	The United Kingdom	Perceptions	470	Quantitative: questionnaire
Cryer and Home (2008)	The United Kingdom	Physical attributes	NA (analysis of literary/ technical materials)	Quantitative: descriptive
International Council on English Braille (1998a)	Australia	Perceptions	68	Quantitative: questionnaire
International Council on English Braille (1998b)	Canada	Perceptions	40	Quantitative: questionnaire
International Council on English Braille (1998c)	Canada and the United States	Perceptions	212	Quantitative: questionnaire
International Council on English Braille (1998d)	Japan	Perceptions	20	Quantitative: questionnaire

Table 5.1 Summary of Studies Included in Review—cont'd

Study	Country	Category	# Participants/ Respondents	Design/ Methodology
International Council on English Braille (1998e)	Nigeria	Perceptions	14	Quantitative: questionnaire
International Council on English Braille (1998f)	South Africa	Perceptions	49	Quantitative: questionnaire
International Council on English Braille (1999a)	Australia, Canada, Japan, New Zealand, Nigeria, South Africa, the United Kingdom, the United States	Perceptions	461	Quantitative: questionnaire
International Council on English Braille (1999b)	New Zealand	Perceptions	56	Quantitative: questionnaire
International Council on English Braille (1999c)	The United States	Perceptions	172	Quantitative: questionnaire
International Council on English Braille (2000)	The United Kingdom	Perceptions	19	Quantitative: questionnaire
White (2011a)	Australia and New Zealand	Perceptions	10	Qualitative: interviews
White (2011b)	The United Kingdom	Perceptions	107	Mixed methods: questionnaire and interviews

were papers or reports located on websites of organizations involved in the development, evaluation, or implementation of UEB or that had been presented at international conferences.

Table 5.1 presents the research articles that were located. Of the 25 articles reporting on research of any kind, only nine articles were published in peer-reviewed journals. Of these nine, one dealt with physical attributes (length) of the code, four dealt with perceptions of the code by individuals representing various constituents, and four presented results of research on the readability of the code. The other 16 articles fell into the category of research but were not published in peer-reviewed journals. Of these 16, two dealt with physical attributes of the code and 14 dealt with perceptions of the code.

3.1. Peer-Reviewed Articles

3.1.1 Physical Attributes of the Code

Only one peer-reviewed research study explored the physical attributes of the code. Knowlton and Wetzel (2006) analyzed the lengths of nine text samples in EBAE, Nemeth code, or CBC and the same passages transcribed into UEB. Depending upon the particular subject, the differences in length varied. UEB was lengthier in all cases, but the subject with the largest discrepancy was algebra. When spatial arrangements were necessary, UEB was 21.9% longer, whereas, with linear expressions, UEB was only 1% longer. The authors indicated that more research would be needed to determine if the additional length of UEB for some subject matter would present problems for students. As for literary materials, UEB was only 4% to 7% longer for the three passages compared.

3.1.2 Perceptions of the Code

Bogart and Koenig (2005) collected survey data from 446 participants (braille readers, proofreaders, educators, and transcribers) on their use of braille and their opinions of UEB. Bogart and Koenig's report was based on the results from administering the same survey/questionnaire in eight countries, Canada, the United States, Japan, Australia, New Zealand, Nigeria, South Africa, and the United Kingdom. The ICEB coordinated the surveys with the braille authorities in each country, and each country produced an evaluation report providing tabulated data on how respondents within that country answered questions about whether they agreed or disagreed with various aspects of UEB (International Council on English Braille, 1998a,1998b,1998c,1998d,1998e,1998f, 1999a,1999b,1999c, 2000).

The participants were English speakers and represented both nontechnical (literary code) and technical (mathematics, science, and computer code) readers, although in Australia, only technical readers responded to the survey. As Bogart and Koenig reported, participants favored the UEB principles in general, and most of the proposed new or changed signs, but were less supportive of the proposed omissions. The authors indicated that the nonrandom selection of participants might have biased the results.

Two research studies used focus groups to collect perceptions about UEB. Wetzel and Knowlton (2006a) conducted five focus groups in conjunction with written responses from individual participants to a questionnaire. Participants were professionals (teachers of children and rehabilitation teachers, braille transcribers, and proofreaders) and consumers (braille readers). The questionnaire and focus groups addressed the potential effects of adopting UEB on users and producers of braille. In general, participants felt that the changes to literary braille would be minor, but the changes to mathematics would be more profound, resulting in passages of much greater length. They expressed concern over whether high school students who were currently using the Nemeth code would be able to become fluent in the new code. Geographic representation was limited, with most participants coming from Minnesota.

Gerber and Smith (2006) studied the results of 13 focus groups throughout Canada. Moderators of the focus groups were teachers trained by the researchers for this role. Each led a focus group of teachers and another focus group of students who were recruited by their teachers. In all, 49 teachers and 36 students participated in one of the 13 different groups. Each focus group explored participants' perceptions of the "barriers and benefits" of UEB, and discussions were audio-recorded. Researchers analyzed the focus group transcripts for negative and positive themes. Students' comments concerning UEB for mathematics were mixed. In general, students' negative comments appeared to reflect a resistance to change, particularly evident for those who used the Nemeth code successfully. Positive comments related to the correlation of UEB to print in that UEB might facilitate better interaction between students who used braille and their peers and teachers in integrated settings. Positive aspects of UEB identified by students included not having to learn CBC and the possibility of increased access to mathematics, science, and computing since students would not be required to learn separate codes. Teachers felt that UEB would be easier to teach and that the changes would help younger readers, struggling readers, beginning computer users, adventitiously blinded individuals, students with multiple disabilities, and sighted parents and teachers learn braille. Participants also gave some directions for future research including the following: measuring actual reading rates for the various codes, examining the ease of learning with respect to time needed to learn and whether people could learn without instruction, and specifying code issues such as differences in space count and whether double cells are a barrier to learning.

Cryer et al. (2013b) reported on a study involving literary braille materials with 27 individuals who read samplers of UEB materials and participated in one of four focus groups or provided individual comments. Participants included students, adult braille readers, teachers, and transcribers. Out of 20 braille readers, five stated that they were not interested in learning UEB. While the respondents were able to read the literary materials without any symbol lists, they expressed a variety of pros and cons of the code and advantages and disadvantages for the adoption of UEB. Transition to UEB was a main concern as was the amount of space needed for materials brailled in UEB. The authors concluded that it would be critical to include key stakeholders in transition planning for UEB if it were to be adopted. Some key points participants felt would make for a smooth transition included the following: decision making regarding the timing of transition for students, the importance of keeping up communication with stakeholders, the development of teaching and reference materials, training for all involved, software updates, liaison with producers including those producing packaging and signage, and the replacement of resources currently available.

3.1.3 Readability of the Code

The articles that examined actual reading of the code were difficult to compare as they studied different aspects of reading in different ways. Wetzel and Knowlton (2006b) used the technique of using cells per second (initiated by Legge, Madison, & Mansfield, 1999) rather than words per minute. Fifty-two (52) participants read cards, each containing three lines of braille consisting of approximately 20 cells per line for literary and mathematical text. For literary text, there was a pattern of longer reading time in cells per second in UEB rather than EBAE when passages examining the use (EBAE) or non-use (UEB) of certain contractions were compared; however, the authors stated that the pattern was not clear, possibly due to the length of sentences or shortness of passages that were tightly controlled. One notable finding was that when reading numbers, there was no significant difference in reading rates for Nemeth (lower) numbers and EBAE (upper) numbers, which are also used in UEB, with two exceptions. Single digits and mixed digits (single, double, and triple digits mixed randomly) were read significantly faster as EBAE (upper) numbers.

Steinman et al. (2004) studied eight adult readers of EBAE to determine the impact UEB would have on their reading rates, oral miscues, and regression rates while reading literary materials only. A pretest/posttest design was used to provide reading rates for EBAE (pretest) and UEB (posttest). Reading passages came from the Nelson–Denny Reading Test using different forms for EBAE (Form G) and UEB (Form H). During the readings, videotapes captured participants' finger movements (regressions) and were also used for oral reading miscues and reading rates. Between the pretest and posttest (a period of about two weeks), participants were provided with the UEB Sampler 1 containing examples of reading materials in UEB and were asked to practice reading the samples to familiarize themselves with UEB. On average, participants read significantly faster with EBAE than with UEB, and the number of regressions was significantly higher for UEB. However, the mean oral miscue rate was significantly lower for UEB than for EBAE, demonstrating that the passages in UEB were read more accurately. The authors speculated that, while slowing them down, the extra vigilance readers required to read UEB contributed to this greater accuracy in reading. The sample size was small, and there was no control for the extent to which participants actually studied the sampler they were provided. The authors anticipated that with more exposure, experienced readers could improve their automatic recognition of the newer elements of the code, and beginning readers would not likely have the same difficulty, as they would be learning the code without any comparison.

Holbrook and MacCuspie (2010) reported on research conducted in the area of mathematics and science with highly expert braille users to determine their perspectives on UEB, which might then influence decisions with respect to the adoption of the code in Canada. The selection criteria for participants were that they were (a) experienced braille readers; (b) employed in high-level mathematics, science, and computer science positions; and (c) not familiar with UEB. This purposive sampling procedure netted five subjects who participated in two days of research. The participants were asked to engage in four UEB-related tasks. The first three were independent tasks, completed with a member of the research team present. For the first task, subjects (without any instruction in UEB) were audiotaped while they read nine samples of materials prepared in UEB from high school science textbooks. They were asked to use a think-aloud technique while they read to provide researchers with information about how they tackled reading unfamiliar symbols. At the end of each sample passage, they were provided with a list of the UEB symbols included in that passage and their meanings. The second task involved listening to a short UEB instructional tape while examining accompanying materials provided in UEB. Task three required them to examine a work passage obtained (without their knowledge) from their employer and to read this using the think-aloud technique while being audiotaped. The fourth and final task was to participate in a focus group with the other participants. Participants felt that UEB represented technical material and other codes and that it was understandable. They improved in their reading of the code with practice and saw no reason why the code could not provide braille readers with access to advanced technical materials.

Cryer et al. (2013a) performed a partial replication of the Holbrook and MacCuspie (2010) study with technical experts involving the same physics, computer science, and mathematics materials used by Holbrook and MacCuspie. Six technical experts in these subjects were selected to review the samples, and, in general, the research followed the same procedures used by Holbrook and MacCuspie. Participants were able to read the braille presented and had a range of opinions, pros and cons, regarding the adoption of UEB. The authors stated that, overall, participants were in favor of moving toward UEB.

3.2. Non-Peer-Reviewed Materials

The other articles that were identified fell into the category of research, but were not published in peer-reviewed journals. However, these reports are important in the context of adoption of UEB and thus were included in the list of 25 research articles located (see Table 5.1). None of the non-peer-reviewed materials contained information on readability.

3.2.1 Physical Attributes of the Code

Bogart et al. (2004) analyzed how frequently numbers were followed by either punctuation or letters. This study replicated an earlier effort to compare the implications of using different number systems (Bogart et al., 2000). UEB code developers had recommended the use of upper numbers, which were previously used only in a literary context, rather than lower numbers, which had been used in the Nemeth code for mathematics. The use of upper numbers for mathematics was an issue of concern for many who had been using the Nemeth code. Because of this controversy, Bogart et al. (2000) utilized a computer program designed to count occurrences of number/letter and number/punctuation combinations and analyzed the results that came from applying the program to 16 textbooks (8429 pages of text including approximately half technical material that would have required the Nemeth code and half that would not). In EBAE and UEB, when a letter follows a number, the use of a letter sign is required to differentiate them from each other since both numbers and letters are formed in the upper part of the cell. Punctuation (formed in the lower part of the cell) cannot be confused with the upper numbers of EBAE or UEB, so there is no need for a punctuation indicator. In the Nemeth code, when a letter follows a number, there is no need for a letter sign since Nemeth numbers (formed in the lower part of the cell) cannot be confused with letters; however, when punctuation follows a Nemeth number, a punctuation indicator is required to differentiate them since both are lower-cell symbols. Thus, depending upon the occurrence of these number/letter or number/punctuation combinations, the authors postulated it would be possible to determine which code would require more symbols to express them. The results of their count indicated that number/punctuation combinations were much more frequent in all texts than number/letter combinations. Applying the appropriate rules of EBAE or the Nemeth code to the passages, there was a need for 499 letter signs when the upper numbers were used and a need for 20,735 punctuation indicators in the Nemeth passages. When UEB rules were applied, 212 letter signs were needed, and no punctuation signs. The authors felt that this reinforced the choice of upper numbers for UEB. One subject for which no materials were prepared was algebra. As Knowlton and Wetzel (2006) discovered later, length was impacted by UEB in algebra more than any other subject.

A study by Cryer and Home (2008) of British Braille (Standard English Braille—SEB) in the United Kingdom evaluated literary and technical materials to determine the impact of UEB on the length of the materials. Literary braille materials were produced in capitalized and uncapitalized SEB. Unlike EBAE where dot 6 is used as a capital indicator and capitalization is shown as in print, SEB capitalization is not typically indicated, although the capability to produce SEB with capitals is an option. UEB was developed to require capitalization indicators similar to EBAE but other than that resembled SEB more closely than EBAE, particularly with respect to the use of contractions crossing syllables. Cryer and Home (2008) showed that for literary materials, UEB increased the length of the passages, but the increased length was due in large part to capitalization. In the publications including technical materials, UEB increased the physical size of the publications, but the extent to which this occurred varied depending upon the material's complexity.

3.2.2 Perceptions of the Code

Prior to 2004, when the code was officially approved by ICEB as ready for adoption, sets of samplers of materials were transcribed into UEB. Some of the research studies reported earlier in the text utilized these samplers or portions of them (Steinman et al., 2004; Wetzel & Knowlton, 2006a), and in the various ICEB countries, the samplers were used to solicit opinions from braille readers, teachers, and transcribers. Braille Authority of North America (2003) distributed the samplers widely and reported on the results of a survey that was completed by 248 individuals. The questions accompanying the samplers asked respondents to identify their feelings toward unifying the code before and after reading the samplers, what they liked or disliked about the new code, and what issues they felt the BANA Board should consider in making a decision regarding the adoption of the code. Overall, respondents were more negative toward the code than positive, even with questions worded to solicit both types of input. They asked the BANA Board to consider asking those using both old and new codes for their opinions, to consider space saving in the code, and to consider the burden on blind readers who would need reeducation.

The Braille Authority of the United Kingdom (BAUK) conducted evaluations in 2007 and 2008 that were reported on their website (Braille Authority of the United Kingdom, 2008). They provided braille readers and others with the samplers that were created in the United Kingdom and obtained their opinions on the samplers and on whether BAUK should adopt UEB. Of the 458 respondents who answered the question as to whether BAUK should adopt UEB, 76% were not in favor, and of the 459 who answered the question as to whether adoption of UEB would benefit future braille readers in the United Kingdom and worldwide, 67% answered in the negative. BAUK determined from their results that it would not adopt UEB at that time but would wait until 2013, but in fact, the United Kingdom adopted UEB in 2011.

White (2011a) interviewed ten braille readers, eight from Australia and two from New Zealand, after both countries were well into the implementation of UEB, to help inform the United Kingdom in their decision as to whether or not to adopt UEB. The ten participants were volunteers attending the Australian Braille Authority (ABA) Annual General Meeting (AGM) in May 2011 and a Round Table Conference that followed the AGM. These individuals, therefore, were more engaged with braille and UEB adoption and implementation than typical braille readers. Participants were interviewed individually with a set of questions regarding UEB and implementation in their home country. While they had mixed opinions on UEB, nine out of ten urged the United Kingdom to adopt UEB for the benefit of future braille readers.

In another survey, White (2011b) evaluated participants' responses to an anthology of essays sent to them as recipients of a monthly magazine in braille produced by RNIB. The essays were about the importance of braille and were compiled from those written by winners of a global essay competition. Of the 165 available recipients, 107 were contacted by phone and answered questions posed by the researchers about what they had read. Those who had heard about UEB comprised 73% of the participants, while 27% had not known about it. They were asked how difficult the essays were to read: 38% indicated that they had read it without difficulty and an additional 49.5% had read it but more slowly, and 12.6% had read it but found it difficult. No participants indicated that they were unable to read the UEB materials. Most respondents commented that they had found reading UEB easier than they had expected. The majority of comments were concerns about the amount of additional space taken up by UEB.

3.2.3 Implementation

In addition to contributing to the knowledge base on the code, many reports contributed information on the need for additional research as UEB is implemented around the world. At the ICEB Fifth General Assembly in 2012 in Johannesburg, South Africa, the country reports included results of implementation in countries that had adopted UEB, which included all but the United States at the time. The report of the Code Maintenance Committee (Landon, 2012) provided for the need for continued maintenance of the code to reflect any changes in print that might occur and to assure that the code did not present any difficulties in translation. Six of the seven country reports included information on the implementation of UEB, although the information was not collected in any standardized manner.

Australia had adopted UEB in May of 2005. By 2012, the country was reporting a full transition to UEB (Simpson, 2012). The country report included information about the development of manuals for transcription, including one for technical materials, and a manual that could be used by software developers and a process for providing licensure for transcribers. Many of these manuals were then used by other countries whose implementation followed Australia's. Holloway (2012) reported on Vision Australia's development of a system for rapid braille transcription using DAISY books.

New Zealand adopted UEB in November 2005 (Reynolds, Bellamy, Stevens, & Smith, 2012). The implementation of UEB in New Zealand "went live" on 1 January 2008 (Braille Authority of New Zealand Aotearoa Trust, 2012). Reynolds et al. (2012) reported on what they considered to be a successful transition to UEB involving implementation from 2008 to 2012 from the standpoint of education, production and proofreading, and libraries. From the standpoint of education, implementation involved professional development activities around the publication of "The Hitchhiker's Guide to UEB" (UEB Curricula Support Writing Group, Blind and Low Vision Education Network NZ, 2007) developed for teachers and their students for help with literary and numerical materials. IEP meetings were used to inform parents, and decisions as to whether a student would continue to use the Nemeth code or transition to UEB for numeracy were determined on an individual basis at IEP meetings.

Producers and libraries faced challenges including an easily solved challenge of how to braille the Maori language in UEB. Production was facilitated by the publication of the *Unified English Braille Manual, New Zealand Edition* (Royal New Zealand Foundation of the Blind, 2011), based upon the Australian edition already completed, and New Zealand and Australia collaborated on the Trans-Tasman Certificate for transcribers and the Duxbury Braille Translator Producers' Manual. The development of a reading program, *Simply Touch and Read* (*STAR*), for adults (Royal New Zealand Foundation of the Blind, 2009) contributed to their ease in learning UEB. The report detailed the types of challenges faced by transcribers and producers and would be an important resource for any country implementing UEB.

Canada had just recently adopted UEB in 2010 and was the fifth of the seven ICEB countries to do so (Goulden, 2012). The country report included information about the issues faced in Canada that made the implementation of UEB more complicated and concerns by producers about the costs associated with the change. Implementation had not yet begun at the time of the report.

Nigeria adopted UEB in 2005 and reported on beginning its implementation of UEB production in earnest in 2009 (Obi, 2012), which involved the use of a booklet prepared by Canada on UEB entitled *UEB, New Symbols and New Usage* produced for use in workshops and the *Unified English Braille Primer, Australian Edition* (Howse, 2008). The report included the country's gratitude to organizations in Canada, Australia, and the United States for provision of materials and supplies and books (Obi).

South Africa reported on their Unified Braille Code (de Klerk, 2012). They no longer call their code UEB because there are 11 official languages in South Africa, all of which were now unified based upon UEB. In South Africa, literary materials were produced in UEB for all school grades after 2008, and the technical aspects of the code were introduced only in the first three grades initially. Teacher training, particularly for technical materials, was still a challenge.

The United Kingdom adopted UEB in October 2011 and reported that they had a draft implementation plan under review (Osborne, 2012). The United States had not adopted UEB at the time of the ICEB Fifth General Assembly in Johannesburg; therefore, the report did not contain any information about implementation.

At the ICEB Fifth General Assembly in 2012, several resolutions were presented regarding the implementation of UEB. Those approved by the conference included resolutions to continue to conduct research into the results of implementation on the perspectives and recommendations of teachers of students who are blind or visually impaired relevant to the implementation of UEB; the perspectives and recommendations of students who use braille relevant to their learning and use of UEB; the perspectives and recommendations of adult braille users relevant to their learning and use of UEB for implementing UEB; and the perspectives and recommendations of individuals responsible for implementing UEB within their respective countries.

The information provided in the research and reports included in this chapter, particularly those from countries that had previously used the Nemeth code, and the continued implementation of UEB in these countries will inform the United States in its implementation efforts. While the work that has been done thus far is moving UEB forward, there are many questions unanswered by the limited research available. These will form the basis of the research agenda for the implementation of UEB in the United States.

4. DISCUSSION

Our goal was to review the current body of available literature on UEB; however, some limitations must be acknowledged. The small number of research studies on UEB and the wide variety of topics, both peer-reviewed and non-peer-reviewed, makes direct comparisons somewhat difficult. This narrative review, meant to provide information about UEB implementation worldwide and provide guidance for the implementation of UEB in the United States, did not include a systematic review of the methodological quality of included articles. In addition, 16 of the 25 (64%) included articles were non-peer-reviewed literature. Despite these limitations, the literature provides a starting point for the development of a research agenda for the United States regarding UEB.

4.1. Physical Attributes of the Code

Much of the research came from concerns over controversial aspects of the code. Some of the articles regarding the physical aspects of UEB were necessarily tied to the development of the code prior to its adoption. Early in the project, the decision was made to base UEB on the current literary code, the most widely used code among English-speaking countries. However, basic decisions had to be made about the symbol set used and the rules governing the use of those symbols. Decisions had to be made about contractions that are the most contextual and thus most ambiguous. Those symbols could be deleted from the unified code, but the implications of each potential deletion were thoroughly discussed. The major goal was to create a unified system with the greatest likelihood of acceptance by average braille reader that was unambiguous, computable, backtranslatable, extensible (i.e., easy to add new symbols as needed), and accurately reflected the original text (Bogart et al., 2000).

The system that would be used for numbers and mathematical content was widely discussed and debated, as mentioned earlier. A replication of the study that looked at letter/number and letter/punctuation combinations in technical material added strength to the contention that upper numbers would be more advantageous to the unified code and cause less disruption to the "base code," literary braille (Bogart et al., 2004). The results of this small study did not convince opponents of the superiority of using upper numbers in UEB (Nemeth, 2004).

It was not only mathematics that caused debate during code development. The rules governing contraction usage, short-form words, and the addition of extra symbols not currently in the basic literary code were also deliberated. Since the project was international, different countries had various rules and guidelines for contraction use, and these differences were discussed openly and at length. (These archived discussions can still be found on the ICEB website, www.iceb.org.) It is not surprising that a number of reports were generated and disseminated prior to 2004 regarding technical details of the code. Survey feedback from potential users of UEB (Bogart &

Koenig, 2005) was an important part of this process and was used to create the final version of the code. Several of the changes to the code were not adopted based on the strongly worded comments from readers of the proposed code (Bogart & Koenig, 2005; Braille Authority of North America, 2003).

Studies examining the physical attributes of the proposed code focused primarily on the length of the transcription. Historically, a number of rules were introduced in literary braille to save space, which would save paper and, presumably, make braille faster to read. UEB eliminated eight contractions (ally, ation, ble, by, com, dd, into, and to) plus one short-form word (o'clock) used in EBAE and also follows print for spacing. The current code allows for some contractions to be used unspaced from the word that follows them (e.g., in the sequence "and for the present," the words "and for the" would follow each other without spaces between). Knowlton and Wetzel (2006) study looked specifically at the length of text samples produced in EBAE and in UEB and found that UEB samples were 4–7% longer than EBAE samples, but no clear pattern emerged as far as longer reading time for these short literary passages. Mathematics expressions written in UEB are generally longer than those written in the Nemeth code, which has been the cause of much concern in the United States (Nemeth, 2004). Similarly, Cryer and Home (2008) compared materials transcribed into UEB to the same articles in SEB (capitalized and without capitals). They concluded that materials in UEB are generally longer, but the amount of increase depended on the complexity of the material. It is difficult to apply the results of this study in the United States, because of the difference between EBAE and SEB in the use of contractions for syllable bridging and because the United Kingdom did not use the Nemeth code for technical materials. It seems clear that the use of UEB will result in transcriptions of greater length to some degree; the effect of these longer transcriptions on reading is, however, unclear.

While there may be some interest in research related to the physical attributes of the code, it is anticipated that these may only relate to the use or nonuse of the Nemeth code. The physical attributes related to literary braille are simply not that different and were discussed at length during the code development process. The United States has opted to maintain the Nemeth code, whereas the other ICEB members have not. Some future research into comparisons across countries of the physical attributes related to readability in Nemeth code and UEB may be of interest to those who wish to propose either a full use of UEB or continued maintenance of the Nemeth code.

4.2. Perceptions of the Code

A great deal of the literature examines opinions and attitudes toward the adoption of UEB rather than objective data on its usefulness. Opposition to change in the United States and in the United Kingdom is well documented (Bogart & Koenig, 2005; Braille Authority of North America, 2003; Cryer et al., 2013b; Nemeth, 2004). Readers in the United Kingdom particularly questioned the use of capitalization in UEB and the additional length of brailled materials. Capitalization is a relatively recent change to SEB, and materials are available without it. Readers in the United States, who read the distributed samples and responded to surveys about UEB, had the perception that the current braille codes did not need to change. BANA's document *The Evolution of Braille* (Braille Authority of North America, n.d.) was written in an effort to explain the rationale behind the changes in the code to adult braille readers and to ameliorate their concerns in hopes that it would lead to greater acceptance over time. Documenting the changing perceptions of UEB among current braille readers and the strategies and actions that were most effective in building positive attitudes about the new code should be part of a research agenda once UEB is implemented in the United States.

4.3. Readability of the Code

The studies on readability of UEB (Cryer et al., 2013a; Steinman et al., 2004; White, 2011b) examined the issue in different ways. Steinman et al. (2004) concluded that findings of slower reading speeds but higher levels of accuracy might have been due to the readers' inexperience with the code and not inherent differences between the current code and UEB. Children who grow up using UEB should not have the same difficulties, as they would have only learned one code. White (2011b) reported that 87.5% of readers surveyed stated they could read materials in UEB, but many reported reading it slower. Cryer et al. (2013b) reported that respondents could read UEB materials, but were still concerned about the amount of space it took and about how a transition to the new code would take place. The additional space mentioned by readers in Cryer et al.'s study could be attributed in part to capitalization. Concerns about the transition seem less related to readability and more about attitudes toward UEB adoption in general.

As mentioned in the preceding text with respect to physical attributes of the code, the implications for future research of the limited research on readability are most likely to be limited to mathematics. However, ongoing studies examining reading rates of braille readers could certainly be done to provide a gross measure of whether rates are being maintained.

4.4. Mathematics and the Code

Much of the opposition to UEB in the United States has been related to the way mathematics is rendered in UEB (Nemeth, 2004). The United States currently uses two numbering systems: The so-called literary numbers in EBAE are the first 10 letters of the alphabet preceded by a number sign, and the Nemeth code numbers are formed in the lower part of the cell. UEB uses numbers in the upper part of the cell in all contexts; the same numbers and symbols are used in both general reading materials (e.g., weather reports and sports scores) and technical materials. The dream of a unified code was that a single set of symbols, characters, and rules would be used regardless of the context.

It is notable that opposition to UEB use for mathematics focuses primarily on the length of mathematical expressions rather than its readability. Gerrey (2004), a scientist who is blind, observed that he found UEB mathematics easier to read than Nemeth code because of its "roomy feel" (p. 69). Holbrook and MacCuspie (2010) and Cryer et al. (2013a), respectively, conducted qualitative studies with expert users of technical codes in Nemeth for the readers in the United States and Canada and British mathematics code for the study in the United Kingdom. Both studies concluded that these users, previously neutral about the new code, found UEB to be usable and understandable, which would not impede their work in technical fields. The country reports do not indicate any difficulty with UEB for mathematics. In fact, Gentle, Steer, and Howse (2012) reported that third and fifth grade students in Australia after only four months of using UEB chose to take their standardized exams in reading and mathematics in the new code. Research on how UEB mathematics is being used in other countries will inform us of whether the need to maintain the Nemeth code is a necessity or simply a desire of those who love the code and its creator, Dr. Abraham Nemeth.

While the United States is committed to maintaining the Nemeth code in addition to UEB, some educators have already expressed interest in using UEB with younger students or with students who have additional learning challenges (Gerber & Smith, 2006). A number of interesting research questions could arise. For example, is the use of a single-number system beneficial to young children rather than learning two different sets of numbers and symbols? If so, when would students transition to Nemeth code and how could this transition be done? New Zealand reported making decisions about Nemeth or UEB code usage on an individual basis through the IEP process; would this be the best approach to use in the United States as well? Do teachers find UEB easier to learn and therefore begin mathematics instruction earlier? Do the length and number of symbols have an impact on reading and understanding mathematical expressions? Questions such as these could lead to important studies that not only would add some much-needed research to our literature base but also would inform other Nemeth code-using countries that are contemplating UEB adoption (such as India).

4.5. Implementation of the Code

Much of the implementation literature from the ICEB country reports could be quite useful to the United States, both in planning its own implementation of UEB and in setting a research agenda. Reports of increased automation of transcription for library collections (Holloway, 2012) lead to speculation about similar processes that could be used in the United States, which would speed up the production of braille books. Resources such as the UEB primers (Howse, 2008; Royal New Zealand Foundation of the Blind, 2011) helped current teachers in Australia and New Zealand learn UEB and could also be adapted for American educators. In fact, they are already available electronically on the BANA website. Programs for teaching braille to adults such as the STAR program in New Zealand (Royal New Zealand Foundation of the Blind, 2009) could be adapted for use in the United States. Canada has already developed an online course for current transcribers who wish to learn UEB and details of similar efforts could be obtained from other ICEB countries. The United States does not have to start from scratch as it transitions to UEB but can build on existing efforts and document the results during implementation.

The United States is also helped in its implementation of UEB by the fact that it is already an international code and thus already compatible with existing assistive technology (AT). Over the next few years, developers of braille technology who have not incorporated UEB into their devices will need to do so to remain viable in a truly global market. UEB is already available in and compatible with most portable braille devices, most notably with the Apple iOS and the Bluetooth braille displays that are used with Apple devices. The availability of UEB in already existing devices increases access to materials in the code so that readers can become familiar with it. Studies could be developed to investigate the role of AT in UEB instruction and dissemination and whether this increased familiarity with the code changes attitudes toward its use.

However helpful it will be to examine implementation reports from other countries, there are important differences to consider. The United States is the only country to adopt UEB and maintain its previously used mathematics code as well. The implications of that decision pose some serious questions related not only to the education of children as mentioned earlier in the text but also for the preparation of teachers. Before students can be taught to read and write UEB, teachers need to be proficient in the code. Country reports from Australia and New Zealand stated the importance of teacher training toward the overall success of UEB implementation. University programs that prepare preservice teachers of students with visual impairments will need to revise their curricula but many questions remain. For how long will preservice teachers need to learn both EBAE and UEB? Will teachers in the United States learn how to teach mathematics with UEB and with the Nemeth code, or will they only learn Nemeth code? How will current teachers be provided with professional development and in-service instruction in UEB? What models of dissemination are most effective for sharing resources and providing training?

Another important difference to note is that many ICEB countries that have adopted UEB have much smaller populations of braille readers than the United States. For example, New Zealand, which began UEB implementation in 2006, currently has 85 school-age students who read braille (Braille Authority of New Zealand Aotearoa Trust, 2012). This contrasts to the approximately 5200 students using braille in the United States (American Printing House for the Blind, 2012). Accurate statistics of the total numbers of braille readers in the United States are difficult to ascertain, but just by virtue of having a larger overall population, one can assume that there are more braille readers in the United States than there are in other ICEB countries. What will be the most efficient way to implement UEB in a country as large as the United States? Some countries started teaching UEB to their youngest students while simultaneously adding UEB materials to library materials for adult readers. Would this model work in the United States? Alternatively, would creating pilot projects in selected states be a better way to approach the implementation? This decision needs to be made with the input of braille readers, educators, and producers so that a coordinated effort prevents another "war of the dots" of wasted resources and time.

Particularly helpful for the United States would be student outcome data from countries that adopted UEB. Unfortunately, to date, nothing has been published from any ICEB country that documents changes in test scores, high school completion, or other data that illustrate the impact of UEB on student performance. Has the adoption of UEB led to an increase in the number of students using braille in ICEB countries? What, then, will be the effect of UEB instruction in the United States on student success? Collecting data on student outcomes as UEB is implemented and over time as the code becomes established could provide some answers.

Other important facets of implementation such as braille production, training, and certification of transcribers will also need attention. For students to learn UEB and teachers to teach it, there must be accurate textbooks produced by well-trained transcribers. Decisions regarding the certification of new transcribers and possible recertification of existing transcribers rest with the National Library Service for the Blind and Physically Handicapped (NLS). Transcribers work in a variety of settings, from schools, to instructional materials centers, to prisons, to home offices. How will current transcribers be trained in the new code and what methods are most effective to meet the needs of this geographically diverse population?

From past history, it is obvious that widespread change does not happen quickly. A complete changeover from current codes to UEB could be anticipated to take as long as ten to fifteen years in terms of materials currently in use being phased out. In the past, code change in the United States and the adoption and implementation of UEB in other countries have happened with, at best, anecdotal recording of the process. We have an opportunity to contribute research to the process of code change in the United States. We hope that this chapter will provide researchers in the field of blindness and visual impairment with significant questions for research on the process of implementing and using UEB and that a research agenda for UEB will be developed to inform those who will deal with the inevitable code changes that will occur in the future.

We hope to see studies that examine perceptions of code change and strategies for increasing positive attitudes toward UEB and investigations into effective transition plans on a state and national level. Research is particularly needed regarding mathematics instruction and how educators will teach mathematics in the future. Most of all, we recommend that student outcomes be monitored as UEB is implemented to determine if UEB lives up to its promise as a robust, unambiguous code that is easy to use and easy to learn to ensure student success.

REFERENCES

American Printing House for the Blind. (2012). *Annual report 2012,* Retrieved from http://www.aph.org/annual-reports/2012/.

Bogart, D. (2000). The literary code. In J. Dixon (Ed.), *Braille: Into the next millennium* (pp. 106–119). Washington, DC: National Library Service for the Blind and Physically Handicapped and Friends of Libraries for Blind and Physically Handicapped Individuals in North America.

Bogart, D., Cranmer, T. V., & Sullivan, J. E. (2000). Unifying the Braille codes. In J. Dixon (Ed.), *Braille: Into the next millennium* (pp. 160–181). Washington, DC: National Library Service for the Blind and Physically Handicapped and Friends of Libraries for Blind and Physically Handicapped Individuals in North America.

Bogart, D., D'Andrea, F. M., & Koenig, A. (2004). *A comparison of the frequency of number/punctuation and number/letter combinations in literary and technical materials,* Retrieved from http://www.brailleauthority.org/research-ueb/contentanalysisfinal11-15-04.pdf (November 15).

Bogart, D., & Koenig, A. J. (2005). Selected findings from the first international evaluation of the proposed Unified English Braille code. *Journal of Visual Impairment & Blindness, 99,* 233–238.

Braille Authority of New Zealand Aotearoa Trust. (2012). Country report from New Zealand. In *Paper presented at the International Council on English Braille Fifth General Assembly, Johannesburg, South Africa.*

Braille Authority of North America. (1972). *The Nemeth braille code for mathematics and science notation, 1972 revision.* Louisville, KY: American Printing House for the Blind.

Braille Authority of North America. (2003). *Executive summary: Questionnaire for BANA/UEBC,* Retrieved from http://www.brailleauthority.org/research-ueb/questionnairereport.pdf.

Braille Authority of North America. (n.d). The evolution of braille. Retrieved from http://www.brailleauthority.org/article/evolution_of_braille-full.pdf.

Braille Authority of the United Kingdom. (2008). *Basic analysis of UEB questionnaire results,* Retrieved from http://www.ukaaf.org/bauk.

Catholic Encyclopedia (n.d.). History of education of the blind. Retrieved from http://www.newadvent.org/cathen/05306a.htm.

Cryer, H., & Home, S. (2008). *Comparing standard English Braille and Unified English Braille codes,* Retrieved from http://www.rnib.org.uk/aboutus/Research/reports/reading/Pages/ueb_seb.aspx.

Cryer, H., Home, S., & Morley Wilkins, S. (2013a). Unified English Braille in the United Kingdom: Part 1—Examination by technical expert braille users. *British Journal of Visual Impairment, 31,* 228–237.

Cryer, H., Home, S., & Morley Wilkins, S. (2013b). Unified English Braille in the United Kingdom: Part 2—Examination by literary braille users, braille teachers, and transcribers. *British Journal of Visual Impairment, 31,* 238–247.

D'Andrea, F. M. (2009). From carvings to computers: A history of tactile codes for people who are blind. *The Educator, 21,* 5–9.

de Klerk, C. J. (2012). Country report for South Africa (2008-2012). In *Paper presented at the International Council on English Braille Fifth General Assembly, Johannesburg, South Africa.*

Gentle, F., Steer, M., & Howse, J. (2012). New dots downunder: The implementation of Unified English Braille (UEB) in Australian schools. *British Journal of Visual Impairment, 30,* 197–200.

Gerber, E., & Smith, B. C. (2006). Literacy & controversy: Focus-group data from Canada on proposed changes to the braille code. *Journal of Visual Impairment & Blindness, 100,* 459–470.

Gerrey, B. (2004). A braille-using scientist embraces the Unified English Braille code. *Journal of Visual Impairment & Blindness, 98,* 69.

Goulden, J. (2012). Canadian report for the International Council on English Braille fifth general assembly. In *Paper presented at the International Council on English Braille Fifth General Assembly, Johannesburg, South Africa.*
Grey, C. (2009). Lessons from the development of Computer Braille Code. *Journal of Visual Impairment & Blindness, 103*, 740–742.
Holbrook, M. C., & MacCuspie, P. A. (2010). The Unified English Braille code: Examination by science, mathematics, and computer science technical expert braille readers. *Journal of Visual Impairment & Blindness, 104*, 533–541.
Holloway, L. (2012). Rapid braille: Automated conversions from DAISY to Unified English Braille. In *Paper presented at International Council on English Braille Fifth General Assembly, Johannesburg, South Africa.*
Hooper, M. S. (1957). The Nemeth code: How and why. *The Education of the Blind*, 7(2), 53–56.
Howse, J. (2008). *Unified English Braille primer: Australia edition.* Sydney, Australia: NSW Department of Education and Training.
Illingsworth, W. H. (1910). *History of the education of the blind.* London, England: Sampson Low, Marsten & Company.
International Council on English Braille. (1998a). *UBC evaluation report—Australia,* Retrieved from http://www.iceb.org/ubcevau.html.
International Council on English Braille. (1998b). *UBC evaluation report—Canada,* Retrieved from http://www.iceb.org/ubcevca.html.
International Council on English Braille. (1998c). *UBC evaluation report—Canada and the United States,* Retrieved from http://www.iceb.org/ubcevcu.html.
International Council on English Braille. (1998d). *UBC evaluation report—Japan,* Retrieved from http://www.iceb.org/ubcevjp.html.
International Council on English Braille. (1998e). *UBC evaluation report—Nigeria,* Retrieved from http://www.iceb.org/ubcevni.html.
International Council on English Braille. (1998f). *UBC evaluation report—South Africa,* Retrieved from http://www.iceb.org/ubcevsa.html.
International Council on English Braille. (1999a). *UBC evaluation report—Consolidated worldwide (expanded sample),* Retrieved from http://www.iceb.org/ubcevwr.html.
International Council on English Braille. (1999b). *UBC evaluation report—New Zealand,* Retrieved from http://www.iceb.org/ubcevnz.html.
International Council on English Braille. (1999c). *UBC evaluation report—US,* Retrieved from http://www.iceb.org/ubcevus.html.
International Council on English Braille. (2000). *UBC evaluation report—UK,* Retrieved from http://www.iceb.org/ubcevuk.html.
Irwin, R. B. (1955). The war of the dots. In R. Irwin (Ed.), *As I see it* (pp. 3–56). New York, NY: American Foundation for the Blind.
Kamei-Hannan, C. (2006). Review of research: Neuroscience and the impact of brain plasticity on braille reading. *Journal of Visual Impairment & Blindness, 100*, 397–413.
Knowlton, M., & Wetzel, R. (2006). Analysis of the length of braille texts in English Braille American Edition, the Nemeth code, and computer braille code versus the Unified English Braille code. *Journal of Visual Impairment & Blindness, 100*, 267–274.
Landon, P. (2012). Report of the code maintenance committee. In *Paper presented at the International Council on English Braille Fifth General Assembly, Johannesburg, South Africa.*
Legge, G. E., Madison, C. M., & Mansfield, J. S. (1999). Measuring reading speed with the MNRead test. *Visual Impairment Research, 1*, 133–145.
Lewis, M. (1970). Must visually handicapped students be low achievers in math? *Education of the Visually Handicapped, 2*, 60–61.
Lorimer, P. (2000). Origins of braille. In J. Dixon (Ed.), *Braille: Into the next millennium* (pp. 18–39). Washington, DC: National Library Service for the Blind and Physically

Handicapped and Friends of Libraries for Blind and Physically Handicapped Individuals in North America.

Maxfield, K. (1928). *The blind child and his reading.* New York, NY: American Foundation for the Blind.

Mellor, C. M. (2006). *Louis Braille: A touch of genius.* Boston, MA: National Braille Press.

Nemeth, A. (2004). My take on the status of braille. *Journal of Visual Impairment & Blindness, 98,* 71–73.

New Beacon (1934). Mathematics and the blind student. In *New Beacon: Vol. 8.* (pp. 146–148), Retrieved from http://s22318.tsbvi.edu/mathproject/ch1-sec6.asp

Obi, J. (2012). Country report for Nigeria (2008-2012). In *Paper presented at the International Council on English Braille Fifth General Assembly, Johannesburg, South Africa.*

Osborne, P. (2012). International Council on English Braille report from the United Kingdom 2008-2012. In *Paper presented at the International Council on English Braille Fifth General Assembly, Johannesburg, South Africa.*

Reynolds, J., Bellamy, S., Stevens, M., & Smith, R. (2012). UEB in New Zealand 2008-2012. In *Paper presented at International Council on English Braille Fifth General Assembly, Johannesburg, South Africa.*

Royal National Institute of Blind People. (2012). *Moon tactile reading code,* Retrieved from http://www.rnib.org.uk/professionals/accessibleinformation/accessibleformats/alttactileformats/Pages/moon_format.aspx.

Royal New Zealand Foundation of the Blind. (2009). *Simply Touch and Read (STAR) contracted braille teaching programme.* Auckland, New Zealand: Royal New Zealand Foundation of the Blind.

Royal New Zealand Foundation of the Blind. (2011). *Unified English Braille manual—New Zealand edition.* Auckland, New Zealand: Royal New Zealand Foundation of the Blind.

Simpson, C. (2012). Country report for Australia 2008-2012. In *Paper presented at the International Council on English Braille Fifth General Assembly, Johannesburg, South Africa.*

Steinman, B. A., Kimbrough, B. T., Johnson, F., & LeJeune, B. J. (2004). Transferring standard English braille skills to the Unified English Braille code: A pilot study. *RE:view, 36,* 103–111.

UEB Curricula Support Writing Group, Blind and Low Vision Education Network NZ. (2007). *The Hitchhikers Guide to UEB,* updated 2013.

Wetzel, R., & Knowlton, M. (2006a). Focus group research on the implications of adopting the Unified English Braille code. *Journal of Visual Impairment & Blindness, 100,* 203–211.

Wetzel, R., & Knowlton, M. (2006b). Studies of braille reading rates and implications for the Unified English Braille code. *Journal of Visual Impairment & Blindness, 100,* 275–284.

White, M. (2011a). *Implementation of UEB (Unified English Braille) in Australia and New Zealand—The experiences of a small sample of native Braille users.* Peterborough: RNIB Braille Development Unit, Retrieved from http://www.rnib.org.uk/aboutus/Research/reports/reading/Pages/australia_newzealand.aspx.

White, M. (2011b). *Evaluation of the views of a sample of RNIB magazine subscribers on an anthology of essays produced in UEB.* Peterborough: RNIB Braille Development Unit, Retrieved from http://www.rnib.org.uk/aboutus/research/reports/reading/pages/ueb_evaluation.aspx.

CHAPTER SIX

Reading Comprehension for Braille Readers: An Empirical Framework for Research

Mackenzie E. Savaiano[1], **Donald L. Compton, Deborah D. Hatton**
Department of Special Education, Vanderbilt University, Nashville, Tennessee, USA
[1]Corresponding author: e-mail address: mackenzie.savaiano@gmail.com

Contents

Abstract

The purpose of this chapter was to examine literature on braille reading comprehension to identify instructional strategies used to increase reading comprehension. Only 12 studies met inclusion criteria, highlighting a void in the literature. Three small groups of literature emerged examining (a) factors affecting braille reading comprehension, (b) comparisons between students reading braille and students with normal vision reading print, and (c) instructional strategies to increase braille reading comprehension. These studies were situated in an existing model of reading comprehension to describe the status of reading research for braille readers and provide direction for future research.

International Review of Research in Developmental Disabilities, Volume 46
ISSN 2211-6095
http://dx.doi.org/10.1016/B978-0-12-420039-5.00004-6

1. INTRODUCTION

Teaching children to read is a major goal of education, and there has been a specific focus in the United States, since the passage of the Elementary and Secondary Education Act of 2001, on using evidence-based strategies to teach reading. Considerable resources were committed to research on reading, resulting in the report from the National Reading Panel (NRP) from the National Institute of Child Health and Human Development (NICHD, 2000). The NRP used meta-analytic tools to review the evidence in five skills thought necessary for reading proficiency: (a) phonemic awareness, (b) phonics, (c) fluency, (d) vocabulary, and (e) text comprehension. Reading comprehension is often seen as the ultimate goal of reading instruction. In fact, the NRP wrote, "Comprehension is critically important to development of children's reading skills and therefore their ability to obtain an education" (NICHD, 2000, pp. 4–1).

1.1. Models of Reading Comprehension for Print Readers

It is important to understand what is meant when discussing reading comprehension. The RAND Reading Study Group defined comprehension as, "the process of simultaneously extracting and constructing meaning through interaction and involvement with written language" (Snow, 2002, pp. 11). To illustrate their definition, the RAND Reading Study Group developed a model of reading comprehension positioning reader, text, and activity within a larger sociocultural context. This model acknowledges possible variability in each of these four areas. Of particular importance to this review is variability in readers' abilities and motivations. Deficits in specific areas of cognitive ability or existing knowledge can affect the level of interaction with text and, therefore, affect readers' ability to construct meaning from text. Although the model presents relevant interactions at a broad level and can be thought of as a general heuristic regarding the relationship between the reader and text, this model does not identify specific cognitive processes involved in reading comprehension.

Perfetti, Landi, and Oakhill (2005) generated a cognitive model of reading (see Fig. 6.1). This model recognizes the bidirectional interactive nature of word reading and linguistics in the processing of written text along with the activation of prior knowledge required for comprehension. The model is broadly based on the simple view of reading (Gough & Tunmer, 1986). In this simple view, reading comprehension is defined as the product of

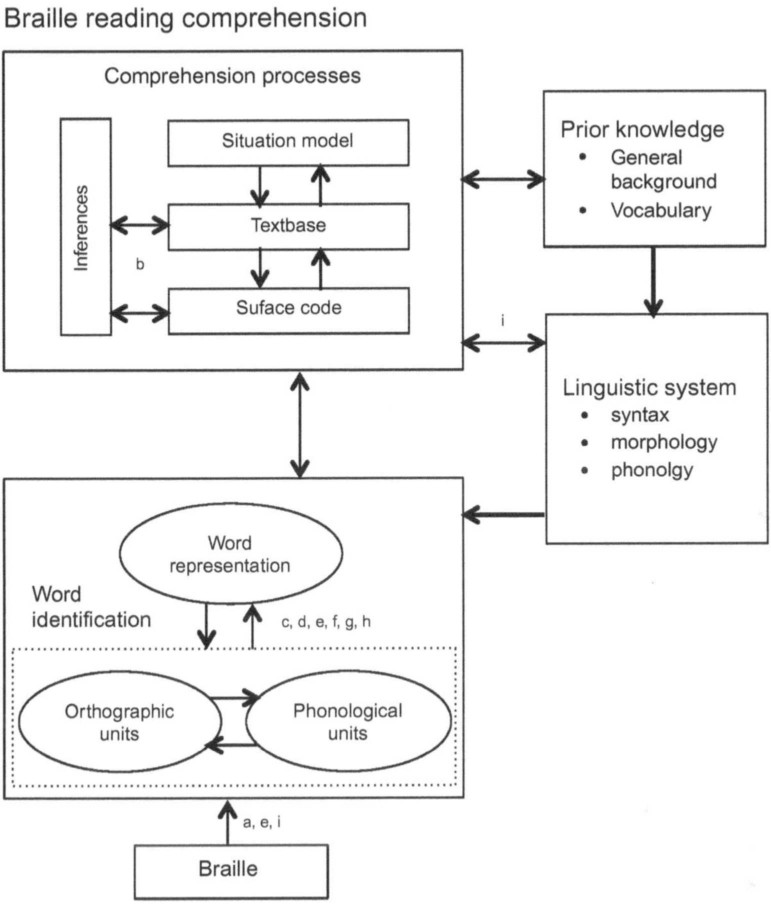

Figure 6.1 Model of reading comprehension. [a]Davidson, Wiles-Kettenmann, Haber, and Appelle (1980), [b]Edmonds and Pring (2006), [c]Harley and Rawls (1970), [d]Hong and Erin (2004), [e]Millar (1987), [f]Millar (1990), [g]Martin and Sheffield (1976), [h]Trent and Truan (1997), [i]Tuttle (1972). *Adapted from Perfetti et al. (2005).*

decoding skills and listening comprehension skills. Perfetti et al. (2005) expanded upon the "simple view of reading" by recognizing the importance of specific knowledge structures (e.g., vocabulary and declarative), subprocesses supporting language processing (e.g., phonology, morphology, and syntax), and the active application of specific processes that support comprehension (e.g., inference making). The model also distinguishes between two different levels of text representations: a text-based representation and a situation model (e.g., Kintsch & Rawson, 2005; Perfetti et al., 2005). The text-based representation conveys the underlying meaning of the text's explicit information, whereas the situation model involves the intertwining of the reader's background knowledge with the text-based representation to form a deep representation of the text. Thus, the situation model is a more meaningful representation that goes beyond the text-based information (Kintsch, 1998), is cumulative, and evolves as one progresses through the text.

The model proposed by Perfetti et al. (2005) is composed of four major components: word identification, comprehension processes, linguistic system, and prior knowledge. Word identification refers to bottom-up processes necessary for decoding and automatic word identification. Comprehension processes are top-down processes involved in creating a global context, or situation model, from which meaning is extracted. The linguistic system reflects knowledge of the English language including vocabulary, syntax, and morphology. Prior knowledge reflects any and all background information readers possess that might influence their efficiency in the other two areas. An important aspect of the model is the interaction among all four systems, represented graphically by double-headed arrows. These interactions are essential (Snow, 2002) and are the reason why deficits in any area can affect the entire process of reading. While there are several differences between reading print and braille, we believe that comprehending text does not differ significant between input systems (print vs. braille). Therefore, we will use the Perfetti model as a heuristic to evaluate the extent to which braille studies have evaluated key factors responsible for both mean differences between typically developing and braille readers and individual differences within students who read braille.

1.2. Braille Reading

Braille is a code that allows children who are blind to access print materials. However, there are significant differences between print and braille. Braille is a unique code in which each character is made from a combination of six dots. There are two forms of braille: uncontracted and contracted. Uncontracted braille assigns a unique dot configuration to each letter in the English alphabet. Contracted braille represents common whole words and common letter combinations with one or two braille cells, removing the one-to-one correspondence between print letters and braille characters. Table 6.1 provides some examples of braille symbols and their meanings. For example, the symbol for the letter *h* can also represent the whole word *have* in contracted braille if it is preceded and followed by a space. If preceded by a dot 5, it can

Table 6.1 Braille Symbols, Meanings, and Dot Configurations

Braille Symbol	Meaning(s)	Dots
⠓	h have 8 (literary code)	1, 2, 5
⠐⠓	here	5 + 1, 2, 5
⠦	his 8 (math code) ? (question mark) “ (opening quotation mark)	2, 3, 6
⠁⠙⠐⠓	adhere	1 + 1, 4, 5 + 5 + 1, 2, 5
⠁⠙⠓⠻⠫	adhered	1 + 1, 4, 5 + 1, 2, 5 + 1, 2, 4, 5, 6 + 1, 2, 4, 6
⠬	ing	3, 4, 6
⠎⠬	sing	2, 3, 4 + 3, 4, 6
⠮	the	2, 3, 4, 6
⠮⠍	them	2, 3, 4, 6 + 1, 3, 4

represent the word *here* on its own or as part of a longer word as in adhere. Being made of only six dots, there are limited configurations for braille characters. As a result, many braille contractions share the same shape but in a different location within the cell. Some braille contractions even share the exact same configuration, requiring syntax and context to discriminate the character. For example, the same configuration is used for the number eight, the word *his*, an opening quotation mark, or a question mark. Meaning depends on the local context.

As the individual characters become more difficult to discriminate, words become harder to read, because the unit of recognition in braille is the individual braille character (Nolan & Kederis, 1969). Reading one character at a time may increase the cognitive load of students who read braille (Steinman, LeJeune, & Kimbrough, 2006). To read a word made up of several cells, braille readers need to identify individual characters consecutively, activating individual phonemic units along the way. This sequential, one cell at a time perception may strain orthographic and phonological processing (Adams, 1990). Sometimes braille contractions match common letter groupings in print. For instance, *ing* is represented by a single braille cell, so when reading the word sing, braille readers can still benefit from the activation of a familiar letter combination. However, braille rules often favor the use of contractions even when jeopardizing correct pronunciation, as in the use of the contraction for *the* in the word *them*. Additionally, spelling patterns that exist in print are not always maintained in braille, as in the differences between *adhere* and *adhered* (see Table 6.1). It is important to highlight these differences, because, as Daneman (1988) confirmed, braille is read character by character instead of in larger units, and this perceptual feature of braille limits the rate at which it can be read.

Reading rates of children who read braille are and have been of major concern in the research literature (e.g., Nolan & Kederis, 1969; Wall Emerson, Holbrook, & D'Andrea, 2009). This research has documented that reading rates of braille readers are approximately half or less of the speed of their age-matched, sighted peers (Lusk & Corn, 2006; Wall Emerson et al., 2009; Wright, Wormsley, & Kamei-Hannan, 2009). Researchers have also described disparities between the reading comprehension ability and chronological age of students who read braille (Dodd & Conn, 2000; Wall Emerson et al., 2009). In fact, the gap between chronological age and reading age in students who are blind is approximately 3 years (Gillon & Young, 2002).

Unfortunately, the body of research on *teaching* reading to children who read braille is limited. Ferrell (2006) conducted a literature review identifying 19 studies of reading interventions for children with visual impairments published from 1963 to 2003; none have been replicated. From these studies, Ferrell distinguished nine promising conclusions regarding literacy instruction, only two of which address braille reading comprehension specifically: comprehension decreases when other stimuli compete for students' attention, and deleting words does not affect comprehension.

Since 2003, there has been more research on braille reading. The Alphabetic Braille and Contracted (ABC) Braille Study was a prospective nonrandomized, 5-year descriptive longitudinal study conducted between 2002 and 2007. Researchers used quantitative and qualitative methods to answer broad questions about literacy development, literacy attitudes, and literacy experiences of children who read braille. From these data, Wall Emerson et al. (2009) found vocabulary, measured by the Brigance (1999), to be highly correlated with overall reading ability, as measured by the Johns Basic Reading Inventory (Johns, 2001). In the last year of the study, 47.2% of the participants were performing below grade level in vocabulary, and, across all years of the study, approximately 60% of the participants were reading below grade level. Descriptive statistics supported the conclusion that participants using uncontracted braille had poorer vocabulary and lower reading levels than participants in similar grades using contracted braille. Although graded word lists and passages were followed by comprehension questions, descriptive comprehension data were reported for 10 of the 42 participants, so the ABC Braille Study (Wall Emerson et al., 2009) did not meet inclusion criteria for this review.

Researchers conducting the ABC Braille Study collected data on subskills of reading and overall reading performance, but not specific strategies used to teach reading. Wormsley (2011) developed an approach to braille reading instruction for children with mild cognitive disabilities called the Individualized Meaning-centered Approach to Braille Literacy Education (I-M-ABLE). This approach focuses on key vocabulary instruction using highly motivating words, phonics, contractions, and teaching efficient hand use. Although there is no experimental research on I-M-ABLE, progress-monitoring data have been published to illustrate the effectiveness of this approach with individual students (Campbell, 2011).

There is a shortage of experimental research on braille reading. However, to move forward with research in this field, we must have a better understanding of what is known about reading. We could not locate a published model of braille reading processes; however, there is extensive research on reading processes in typically developing students and students with learning disabilities. Although there are several key differences between the reading of print and the reading of braille, it seems reasonable at this stage in our understanding of braille reading to learn from research in other disability areas. The purpose of this review was to examine the existing literature to determine what factors impact braille reading comprehension. Consequently, articles identified by this review will be discussed within the context of the model of reading comprehension described by Perfetti et al. (2005).

2. METHOD

2.1. Search Procedure

A search was conducted using Dissertation Abstracts International, ERIC, PsycINFO, Education Full Text, and ProQuest Central databases using the terms "reading comprehension" and "braille" and "low vision" or "vis* impair*" or "partial* sight*" or "blind*" or "functional* blind*."

A substantial number of articles (362) were identified by the search. To be included in this review, research studies must have (a) reported at least one measure of reading comprehension, (b) included school-age children who read braille as participants, and (c) used materials produced in English braille. Studies were excluded if participants were primarily foreign language students, English language learners, learning-disabled, or deaf. After applying these criteria to the search, 69 articles were left. Of the remaining articles, 48 (70%) were from journals of neuroscience and only mentioned blindness and braille as an example of the brain's plasticity. Three (4%) articles were eliminated because some of the participants were not between the ages of 3 and 21 years, nine (13%) articles were eliminated because participants did not read braille, and two (3%) articles were eliminated because they used non-English braille codes. A total of seven (10%) articles returned by the search met all inclusion criteria.

As articles were identified as meeting inclusion criteria, their reference lists were checked to identify further articles of interest. In addition, all articles included in Ferrell's (2006) review were examined, and searches of the following journals (1965–present) were conducted by hand to ensure inclusion of all relevant articles: *British Journal of Visual Impairment*, *Education of the Visually Handicapped*, *Journal of Visual Impairment & Blindness*, *New Outlook for the Blind*, *Reading Research Quarterly*, *RE: view*, *Scientific Studies of Reading*, and *Visual Impairment Research*. Using these methods, three additional articles were identified. A total of 12 studies in 10 articles met criteria for inclusion (see Table 6.2 for a summary of all studies included in this review).

2.2. Quality of Included Articles

Articles meeting criteria for inclusion were examined using quality indictors appropriate to the type of research design. Rubrics based on quality indicators from the Council for Exceptional Children Division for Research for correlational research (Thompson, Diamond, McWilliam, Snyder, & Snyder, 2005) and for single-subject research (Horner et al., 2005) were used to assess articles in this review. A second reader trained in using the rubric coded 100% of the correlational articles independently. Discrepancy discussions occurred whenever there was a disagreement between coders and a consensus code recorded. To retain information from the independent coding, reliability data were collected on the initial coding of the articles, not the consensus code. Raters achieved interobserver agreement of 91%.

Table 6.2 Summary of Studies Included in Review

Study	Design	N	Age/Grade of Participant(s)	Visual Info.	Reading Passage(s)	Comprehension Measure(s)	Data Analysis	Groups	Results
Tactile input (n = 5)									
Davidson et al. (1980)	Corr.	18	Range = 13–21 M = 16.38	<LP	Modified graded readers; approximately 150 words Contracted brl	Oral recall of 10 important facts (% recalled) 10 MCQ (% correct)	4 × 4 ANOVA	1: M = 4.9 grade (n = 5) 2: M = 6.8 grade (n = 5) 3: M = 9.9 grade (n = 4) 4: M = 14.1 grade (n = 4)	No HM × RA interaction Significant RA × PD interaction
Martin and Sheffield (1976)	Corr.	171	7th–12th grade	Not reported	Traditional passage = 947 words 10% telegraphed; 30% telegraphed; 50% telegraphed Contracted brl	Set relations: 20 MCQ (# correct) Passage details: 40 MCQ (# correct)	2 × 4 ANOVA	VI: n = 27 NV: n = 144	No sig. main effect for group Significant group × rate interaction Significant group × time interaction
Millar (1987), Study 1	Corr.	18	M = 17	Not reported	Researcher created stories Contracted brl	8 MCQ (proportion correct)	3 × 2 × 2 ANOVA	1: M = 144 wpm 2: M = 114 wpm 3: M = 74 wpm	Comprehension with vertical FO < comprehension with horizontal FO
Millar (1987), Study 2	Corr.	18	M = 17	Not reported	Researcher created stories Contracted brl	8 MCQ (proportion correct)	3 × 2 × 2 ANOVA	1: M = 144 wpm 2: M = 114 wpm 3: M = 74 wpm	Intact text > degraded text in Groups 1 and 2
Tuttle (1972)	Corr.	104	Range = 14–21	Not reported	Reading Versatility Test, Intermediate Level Brl contraction level not reported	Reading Versatility Test, Intermediate Level	4 × 3 ANOVA	1: High brl, high listening (n = 39) 2: High brl, low listening	Brl > listening in Group 2 Listening > brl in Group 3 Brl > compressed

Continued

Table 6.2 Summary of Studies Included in Review—cont'd

Study	Design	*N*	Age/Grade of Participant(s)	Visual Info.	Reading Passage(s)	Comprehension Measure(s)	Data Analysis	Groups	Results
						24 MCQ (# correct)		($n=13$) 3: Low brl, high listening ($n=13$) 4: Low brl, low listening ($n=39$)	speech in Group 1 Brl less time-efficient for all groups
Word identification (n = 5)									
Harley and Rawls (1970)	Corr.	39	Range = 5–10 $M=7.7$	20/200—NLP	GORT Types of brl: 1: Uncontracted 2: Contracted 3: Phonemic	GORT	*t*-tests	1: Analytic ($n=4$) 1: Synthetic ($n=11$) 2: Analytic ($n=4$) 2: Synthetic ($n=8$) 3: Analytic ($n=8$) 3: Synthetic ($n=4$)	Not all groups comparable on MA Significant group × contraction level interaction
Hong and Erin (2004)	NA	16	Range = 7.75–18.83 (2nd–12th) $M=12.16$	20/800—NLP	Johns Basic Reading Inventory Contracted brl	Johns Basic Reading Inventory 10 MCQ (# incorrect)	NA	Contracted ($n=8$) Uncontracted ($n=8$)	Descriptive: No differences in mean # MCQ incorrect between groups
Millar (1990), Study 1	Corr.	21	Range = 11.0–19.3 $M=14.11$	Not reported	Researcher modified stories; $M=92$ words Contracted brl	8 MCQ (# correct)	3 × 5 ANOVA *t*-tests	1: $M=145$ wpm ($n=7$) 2: $M=70$ wpm ($n=7$) 3: $M=28$ wpm ($n=7$)	Suppression < all other conditions

Millar (1990), Study 2	Corr.	20	Range = 6.4–19.4 *M* = 13.5	Not reported	Researcher modified stories Contracted brl (controlled for easy contractions)	8 MCQ (# correct)	4 × 5 ANOVA *t*-tests	1: *M* = 212 spm (*n* = 5) 2: *M* = 107.8 spm (*n* = 5) 3: *M* = 59.2 spm (*n* = 5) 4: *M* = 19.6 spm (*n* = 5)	Suppression < all other conditions for Group 4 Suppression < silent reading for Group 2
Trent and Truan (1997)	NA	30	Not reported	Not reported	GORT, level 6, Form C or D	GORT, level 6	NA	NA	See Table 6.3
Comprehension processes (n = 1)									
Edmonds and Pring (2006)	Corr.	34	VI (*n* = 17): Range = 7.9–11.7 *M* = 9.11 RM (*n* = 17): Range = 7.5–9.2 *M* = 8.2	NLP—partial sight	Researcher constructed stories, approximately 100 words Uncontracted brl	NARA (comp. age) 8 MCQ: 4 Literal, 4 inference (% incorrect)	2 × 2 × 2 ANOVA *t*-tests	VI/AC: *n* = 9 VI/LAC: *n* = 8 RM/AC: *n* = 7 RM/LAC: *n* = 10	No significant main effects Significant QT × RA interaction
Prior knowledge and motivation (n = 1)									
Creech and Golden (2009)	SS: changing criterion	1	16	Blind with moderate MR	Restaurant menu Included known contractions	5 MCQ (% correct)	Visual analysis	N/A	No functional relation between RC and token system

Note: Corr., correlation; M, mean; LP, light perception; brl, braille; MCQ, multiple-choice questions; HM, hand movement; RA, reading ability; PD, passage difficulty; wpm, words per minute; FO, finger orientation; spm, syllables per minute; VI, visually impaired; GORT, Gilmore Oral Reading Test; RM, reading matched; NLP, no light perception; NARA, Neale Analysis of Reading Ability; AC, advantaged comprehender; LAC, less advantaged comprehender; QT, question type; NV, normal vision; SS, single subject; MR, mental retardation; MA, mental age.

3. RESULTS

The model of reading comprehension adapted from Perfetti et al. (2005) was used as a framework for discussing the articles identified by this review. As such, the research was organized as it relates to areas of the model, specifically the four major components and how they interact with each other.

3.1. Interaction of Braille and Word Identification Processes

Five studies in four articles investigated factors involving braille itself (Davidson et al., 1980; Martin & Sheffield, 1976; Millar, 1987; Tuttle, 1972). Two studies examined possible effects of the mechanics of reading braille, specifically hand movements and finger orientation (Davidson et al., 1980; Millar, 1987, Study 1), and three studies examined possible effects of text quality or mode of input (Martin & Sheffield, 1976; Millar, 1987, Study 2; Tuttle, 1972).

Braille mechanics: Hand patterns refer to the position and movement of the hands when reading braille. Students can use a one-handed pattern (either left or right hand) or a two-handed pattern (left marks, parallel pattern, split pattern, or scissor pattern). Davidson et al. (1980) video-recorded 18 students reading braille passages aloud from least to most difficult (2nd, 4th, 6th, and 12th grade), retelling passages in response to a prompt, and answering ten multiple-choice questions for each passage; videos were coded for type of hand pattern. Davidson et al. (1980) found no association between hand pattern and either measure of reading comprehension when participants were grouped by reading ability (see Table 6.2). There was a significant interaction between reading ability and passage difficulty on comprehension with the better readers scoring significantly higher on both measures of comprehension, even as passages increased in difficulty. There was also a positive correlation between reading ability and IQ, which could have contributed to the significant interaction. All 18 participants made a considerable number of single regresses (moving back to the previous braille character then continuing on), irrespective of reading ability group or passage difficulty.

Millar (1987, Study 1) found that reading braille silently with vertical finger orientation produced poorer reading comprehension than horizontal finger orientation for all 18 participants regardless of their reading rate. Finger orientation corresponds to the direction of the reading finger(s) in relation to the body when reading braille that has been rotated 90°

counterclockwise. This rotation created an unfamiliar reading task in both orientation conditions: familiar body position/unfamiliar text arrangement (vertical) and unfamiliar body position/familiar text arrangement (horizontal).

Text quality: Unlike print, braille can degrade over multiple readings. Millar (1987, Study 2) investigated the effects of braille quality on students' reading comprehension. In this experiment, all 18 participants silently read passages with intact braille and passages with degraded braille for comprehension and letter search tasks. Researchers degraded the braille by smoothing down one dot within a character. In the degraded condition, a total of eight degraded characters were distributed randomly throughout a passage. All groups read the degraded passage slower. Moreover, degraded braille affected performance on the reading for comprehension task. However, comprehension was only affected for Groups 1 and 2, the two faster groups. Millar (1987) explained this effect as a difference in the scanning technique of fluent braille readers. Rather than feeling for a general character shape, Millar hypothesized that fluent braille readers were making rapid judgments about the relative position of dots to gaps within a braille character to determine its meaning. Thus, the degraded braille significantly affected the comprehension of only the fastest readers.

Martin and Sheffield (1976) degraded passages in a different way. They investigated the impact of telegraphic passages on the reading ability of students who read braille and compared their performance to a sample of students with normal vision (NV). Telegraphic in this context referred to a passage abbreviated by deleting nonessential words. This experiment occurred in two phases. In Phase 1, Martin and Sheffield (1976) created the deletion level passages based on the mean rank order for words in each sentence of the full passage. In other words, participants decided which words were least important to the meaning of the story, and a certain percentage (10%, 30%, and 50%) of the least important words were deleted. In Phase 2, a random sample within each group read the full (braille = 7, NV = 36), 10% (braille = 6, NV = 36), 30% (braille = 7, NV = 36), and 50% deletion level passages (braille = 7, NV = 36). In the NV group, better performance on comprehension measures was associated with being in a higher grade, regardless of the deletion level. In the braille group, there was a significant deletion level effect on comprehension, but none of the individual comparisons reached significance. Descriptive statistics suggest poorest performance was on the 50% deleted passage. The braille group read slower and took more time to read than the NV group. There were also significant interactions between group and deletion level for both variables. The average reading rate of the NV group decreased as deletion level increased, and the braille group took less time to read as deletion level increased. There were no significant differences between groups on either measure of comprehension.

Tactile versus auditory input: Speed in reading, as mentioned earlier, is often the major concern for teachers of students with VI. As a result, many braille readers end up reading by listening to audio or digital recordings in an effort to keep up with the content of middle and high school classes. Tuttle (1972) compared the comprehension of students reading braille with comprehension when listening to normal or compressed speech. Compressed speech refers to a sped-up recording (approximately 250 wpm) that does not distort pitch or voice. The 104 participants were grouped by their ability in each mode (see Table 6.2). Some findings resulted from the grouping variable: Group 2 (high braille, low listening) performed significantly better when reading braille, and Group 3 (low braille, high listening) performed significantly better in both listening conditions when compared to reading braille. However, when participants were considered as a whole, there were no significant differences between the three conditions. Tuttle (1972) also created an index of learning efficiency (number of correct answers divided by number of minutes spent reading) and reported that braille was significantly less efficient for all participants, regardless of group.

Tuttle's (1972) index of efficiency may be misleading, however. While efficiency implies maximum productivity with minimum effort, Tuttle's measure appeared to be an index of time efficiency. Considering Groups 1 and 2 performed significantly better in the braille condition, the extra time needed did not indicate inefficiency. Additionally, there appeared to be a bias toward the listening conditions because each was recorded at a speed set by the researchers. In natural environments, students adjust speed settings independently for their preferences. These speed preferences were not addressed in Tuttle's metric.

3.2. Word Identification Level Processes

Five studies in four articles investigated word identification level processes of the model (Harley & Rawls, 1970; Hong & Erin, 2004; Millar, 1990; Trent & Truan, 1997). Two studies examined effects of reader variability on braille reading comprehension (Hong & Erin, 2004; Trent & Truan, 1997), two examined the effects of articulatory suppression on reading comprehension (Millar, 1990), and the other examined differences between two reading curricula when transcribed in different forms of braille (Harley & Rawls, 1970).

Reader variability: In an effort to identify factors affecting braille comprehension, Trent and Truan (1997) grouped students in two ways: by age of onset of blindness and by reading speed. Though Trent and Truan (1997) only reported descriptive statistics for their study, they presented raw data in text, which permitted an analysis by reading speed (see Table 6.3). There was a significant correlation between comprehension and accuracy, but not between comprehension and reading rate. However, there was a significant correlation between accuracy and rate, which implies an indirect association between rate and comprehension. Still, there was no significant difference in comprehension between students grouped by reading speed. Raw data by

Table 6.3 Analysis of Trent and Truan (1997) Comprehension Data

Correlation matrix

	Comprehension	Rate	Accuracy
Comprehension	1.0		
Rate	0.22	1.0	
Accuracy	0.58[a]	0.60[a]	1.0

Descriptive statistics

		Reading Comprehension (%)		
Group		*M*	(SD)	*n*
High	80–103 wpm	81.7	(14.7)	6
Middle	20–79 wpm	77.1	(25.7)	17
Low	8–17 wpm	68.6	(28.5)	7
Total		76.0	(24.3)	30

One-way ANOVA

	Sum of Squares	df	Mean Square	*F*	*p*
Model	598.011[b]	2	299.006	0.489	0.619
Residual	16521.989	27	611.926		
Total	17120.000	29			

[a] $p < 0.001$.
[b] $R^2 = 0.035$.

age of onset of blindness were not reported. The descriptive summary connected late blindness (defined as losing vision after 3rd grade) with slower reading speeds but did not address the comprehension of these groups.

Hong and Erin (2004) grouped 16 braille readers as contracted or uncontracted based on the code used when they learned to read. Reporting descriptive statistics, they noticed little difference between groups' mean comprehension performance.

Articulatory suppression: Millar (1990) conducted two studies on the effect of articulatory suppression (making unrelated speech sounds to limit subvocalization) on the silent reading comprehension of students who read braille. Both studies included a read-aloud condition, to control for related speech sounds, and a foot-tapping condition, to control for nonspeech movement. The first study divided 21 experienced braille readers into groups based on reading rate (see Table 6.2). Overall, articulatory suppression significantly affected the comprehension scores of all three groups, though comprehension did not differ significantly across groups.

The second study not only utilized similar procedures but also included beginning braille readers. The 20 participants were also divided into groups based on reading rate (see Table 6.2). Again, Millar (1990) found that articulatory suppression significantly affected the comprehension scores of all three groups. In addition, the group by condition interaction was significant. Specifically, articulatory suppression elicited significantly worse comprehension than other conditions in the beginning reader group. Groups 1 and 3 did not perform significantly different in any condition. Reading comprehension for Group 2 was significantly better in silent reading conditions compared to articulatory suppression, but students' performance under articulatory suppression was not significantly different from oral reading conditions.

Comparison of reading curricula: Harley and Rawls (1970) investigated two contrasting and commercially available reading curricula, each prepared by the authors in three types of braille: uncontracted, contracted, and phonemic. Teachers were trained to use the different curricula, and the curricula were implemented throughout one school year. Participants' reading comprehension was tested at the end of the year using the Gilmore Oral Reading Test (Gilmore & Gilmore, 1968). Researchers found that the uncontracted group was significantly different from the other two groups with respect to mental age, so their results were not used in the final analysis. Findings showed participants using materials in contracted braille performed better on comprehension measures, regardless of the curriculum. Researchers only reported significant findings, so it is unclear whether the comparison between contracted-analytic and contracted-synthetic approaches was tested.

3.3. Comprehension Level Processes

Only one study examined comprehension level processes of the model. Edmonds and Pring (2006) compared the performance of children who read braille and children with NV, matched for decoding age, on literal and inferential comprehension tasks. Participants were grouped as advantaged comprehenders (braille = 9, NV = 7) or less advantaged comprehenders (braille = 8, NV = 10). This assignment was based on the discrepancy between participants' observed comprehension age and that predicted by their decoding age. The braille group was significantly older than the NV group, so age was entered as a covariate. There were no significant differences between braille and NV groups. In fact, the only significant finding was the interaction between comprehension group and question type; less advantaged comprehenders performed worse on inferential questions than advantaged comprehenders.

Edmonds and Pring (2006) did not find significant differences in comprehension between groups. These nonsignificant findings show that braille readers can comprehend just as well as their peers. However, Edmonds and Pring used age as a covariate because they matched participants on reading age and found that braille readers were significantly older than reading age-matched peers. Braille readers were performing, on average, below expectations for their age.

3.4. Prior Knowledge and Motivation

Creech and Golden (2009) used a single-subject changing criterion design with one participant. Their investigation of whether a token system could increase the number of correct responses to comprehension questions was a follow-up study. The participant had already been exposed to the token system with the goal of increasing the number of words written on a braillewriter. The initial study demonstrated the effect of the token system, but did not replicate across criterion changes. There was no replication of effect in the follow-up study, because the student reached the criterion set for the first intervention phase during the final baseline session. Consequently, there was no evidence of a functional relation between the token system and the participant's increase in correct answers to comprehension questions.

Though there was no demonstration of a functional relation between the token system and comprehension increases, it is possible that the student in Creech and Golden's (2009) investigation generalized the strategy from the initial study. There were measured increases in comprehension; they just could not be linked explicitly to introduction of the token system.

3.5. Research Quality

There is little research available on braille reading comprehension. Only four articles identified in this review were written in the past 20 years (Creech & Golden, 2009; Edmonds & Pring, 2006; Hong & Erin, 2004; Trent & Truan, 1997), with the rest written between 1970 and 1990. In addition to the synthesis of research findings, a rigorous analysis of all the articles was conducted to determine the quality of research on braille reading comprehension. Articles meeting criteria for inclusion were examined using quality indictors appropriate to the type of research design utilized. Rubrics consisting of quality indicators for correlational research (Thompson et al., 2005) and single-subject research (Horner et al., 2005) were used to assess articles in this review. Each quality indicator was scored on a scale from 0 to 3, with a score of 3 indicating that all anchors for a specific indictor had been met. The summary of studies can be seen in Tables 6.4 and 6.5.

Creech and Golden (2009) did not establish a stable baseline, were unable to demonstrate a functional relation, and did not collect procedural fidelity data, impacting the internal validity of their study (see Table 6.4). Though interobserver agreement data were collected on fewer than 25% of sessions, the four sessions for which it was collected reached an agreement of 100%.

The correlational studies had similar strengths and weaknesses (see Table 6.5). All researchers used appropriate methods in the presence of multiple outcome variables. On the other hand, none provided evidence that assumptions for statistical testing were met. One study reported score reliability and validity (Martin & Sheffield, 1976), and one study reported confidence intervals for some statistics (Tuttle, 1972). No study reported effect sizes, but most studies reported the necessary statistics for the reader to calculate effect sizes. While the quality indicators suggested by Thompson et al. (2005) include GLM weights, the studies identified by this review primarily used ANOVAs, not regressions, and both coders agreed that GLM weights were not an appropriate indicator for the quality of these articles.

Table 6.4 Coding Summary for Single-Subject Study Included in Review
Creech and Golden (2009)

Description	
Participants	1
Intervention components	3
Dependent variable	2
Description score	6
Description quality (proportion)	0.67
Internal validity	
Baseline	1
Visual analysis	2
Implementation fidelity	0
IOA	3
% of sessions for IOA	1
Replication of effect	0
Internal validity score	7
Internal validity quality (proportion)	0.39
External validity	
Maintenance	3
Generalization	0
Social validity	2
External validity score	5
External validity quality (proportion)	0.56
Overall score	18
Overall quality (proportion)	0.50

Note: Scores for each indicator are on a scale from 0 to 3, with a score of 3 meaning all anchors for that indicator are met.

Overall, studies met approximately 28% (range = 15–50%) of the quality indicators appropriate to their design (see Tables 6.4 and 6.5). The low incidence of students who read braille is a challenge for research, often limiting the power available to detect significant differences (Wright, 2010).

Table 6.5 Coding Summary for Correlational Studies Included in Review

Quality indicators	Davidson et al. (1980)	Edmonds and Pring (2006)	Harley and Rawls (1970)	Martin and Sheffield (1976)	Millar (1987), Study 1	Millar (1987), Study 2	Millar (1990), Study 1	Millar (1990), Study 2	Tuttle (1972)
Measurement: reliability	0	0	0	1	0	0	0	0	0
Measurement: validity	0	0	0	0	0	0	0	0	0
Reliability/validity and results	0	0	1	0	0	0	0	0	0
Interobserver agreement	0	0	NA	NA	0	0	0	0	0
Effect size reporting[a]	0	0	0	0	0	0	0	0	0
Effect size calculation (*only if ES is not reported*)	3	3	0	3	3	3	3	3	0
GLM weights	NA	NA	NA	NA	NA	NA	NA	NA	NA
Scale conversion	1	1	3	3	1	1	1	1	1
Methods	3	3	3	3	3	3	3	3	3
Post hoc analysis	2	0	0	3	3	2	0	1	0

Testing statistical assumptions	0	0	0	0	0	0	0	0	0
Confidence intervals	0	0	0	0	0	0	0	0	1
Total score	9	7	7	13	10	9	7	8	5
Overall quality (proportion)	0.27	0.21	0.23	0.43	0.30	0.27	0.21	0.24	0.15

[a]Effect size (ES) reporting was a prerequisite for three coding criteria (ES comparison, ES definition, and ES interpretation). None of the included articles reported ES; therefore, those three criteria were not included in the table.

Note: Scores for each indicator are on a scale from 0 to 3, with a score of 3 meaning all anchors for that indicator are met.

Considering that these quality indicators are meant to be rigorous and represent the gold standard for research, this literature falls short of what can be considered high-quality research. However, because of the scarcity of literature on braille reading comprehension, these articles and their findings are relevant despite their shortcomings.

4. DISCUSSION

The majority of research on reading skills of braille readers has focused on the word identification level processes involved in reading comprehension (see Fig. 6.1).

4.1. Interaction of Braille and Word Identification Processes

Davidson et al. (1980) and Millar (1987, Study 1) investigated the interaction of the reader with the text, specifically how tactile input interacts with word recognition to affect readers' overall comprehension. Neither study focused on the braille itself, but rather on the manner in which it was read (i.e., hand pattern and finger orientation).

Most of Davidson et al.'s (1980) participants used a two-handed pattern to read braille. Other research suggests that reading braille with a two-handed pattern is associated with higher proficiency in reading than reading braille with one hand (Kusajima, 1974; Wright et al., 2009). There may not have been enough participants reading with a one-handed pattern to find a significant association between hand movements and comprehension. In addition, the prevalence of single regresses across reading ability groups is consistent with Daneman's (1988) hypothesis that the perceptual features of braille make it more difficult to read. Millar (1987, Study 1) found that horizontal finger orientation was related to higher comprehension scores, which supports current practice for teaching braille hand and finger movements. Overall, these two studies confirm existing braille reading techniques rather than providing direction for research on braille reading instruction.

Millar (1987, Study 2) and Martin and Sheffield (1976) also investigated the interaction of braille and word identification processes and braille by having participants read degraded braille. Millar showed that reading comprehension could be affected by braille quality. All participants read degraded braille slower, and those who would normally be considered faster, more proficient readers were the ones whose comprehension was affected by poor braille quality. Not only does this provide a reminder that braille needs to be stored properly and replaced more often than print, but also Millar's theory implies that beginning braille instruction should focus on dot positions within the cell, and not on general character shapes.

The effects of degraded print on comprehension processes are widely agreed upon. Illegible or poor-quality print slows down the orthographic processor, which slows down the phonological processor (Adams, 1990). The effect of this slower processing is that letters are perceived independently or in smaller groups, requiring more time to process a word and determine its meaning. Braille, however, is normally perceived character by character, so degrading braille may not have the same effect on reading comprehension processes as degrading print (Daneman, 1988; Nolan & Kederis, 1969).

4.2. Word Identification Level Processes

One way the phonological processor compensates for a slowing down of orthographic processing is through subvocalization. This strategy maintains the stimulation of previous letters/characters in a word as it is being read. It is possible this strategy is used by braille readers to compensate for the slow orthographic processing resulting from the single unit of recognition.

Millar (1990) disrupted this compensatory process by suppressing subvocalization. Not surprisingly, this significantly affected comprehension processes. These results have also been found in proficient print readers (e.g., Daneman & Newson, 1992; Slowiaczek & Clifton, 1980), supporting Steinman et al.'s (2006) belief that braille readers and print readers may experience similar stages of reading. Millar's second study was less conclusive; she found that beginning braille readers were affected significantly more than proficient braille readers (1990, Study 2). One interpretation is that beginning readers were differentially affected by suppression: students in Group 4 were unable to read the story when required to make unrelated speech sounds and were, therefore, unable to answer comprehension questions, whereas students in Group 1 did not perform significantly different than in other conditions. This difference between groups suggests that beginning readers are relying on subvocalization more than proficient readers when reading for meaning. An alternate explanation for this finding could be that proficient braille readers have incorporated subvocalization as a fundamental strategy, making it more difficult to suppress. It is also possible that one or more of the five participants in Group 4 did not subvocalize. With such small groups, the mean would have been substantially affected by the performance of one student.

Contracted braille has no English print equivalent, so there is no comparison to print reading when researchers examined differences between uncontracted braille and contracted braille. Although Harley and Rawls (1970) reported that contracted braille was better than uncontracted or phonemic braille materials when using an analytic or synthetic approach to teaching reading, the tests administered at the end of the year were transcribed in contracted braille for the contracted and phonemic groups, possibly providing an advantage to the contracted group. Because the uncontracted group was significantly different from the other two groups, data from the uncontracted group were not used in the analysis. Therefore, the results only compared contracted braille with phonemic braille. Also, by only reporting significant findings, it is not known whether researchers tested whether the contracted-synthetic approach was superior to the contracted-analytic approach.

Hong and Erin (2004) compared students who learned to read using different kinds of braille, and their results seem contrary to the ABC braille study finding that students using uncontracted braille had poorer vocabulary and lower reading levels than participants in similar grades using contracted braille (Wall Emerson et al., 2009). However, all of Hong and Erin's (2004) participants lost their vision before entering school and used contracted braille at the time of the study. Their groups were based on which code they learned first, suggesting that students who read contracted braille perform comparably, as long as they learned to read using braille.

4.3. Comprehension Level Processes

Edmonds and Pring (2006) were interested in whether there are differences in inferencing between students who read braille and students who read print. They did not find differences between groups, but did find that students whose comprehension age was less than that predicted by their decoding ability performed poorer on inferential questions. The comparison to reading age-matched peers demonstrates that a gap still exists between braille readers and their peers, but this study did not attempt to discover whether this gap was significant when comparing braille readers to chronological age-matched peers. This finding reinforces the dynamic comprehension processes represented in Fig. 6.1, since deficits in any one area influence all the other processes.

4.4. Prior Knowledge

Even though the importance of prior knowledge and its interaction with word identification and comprehension processes is highlighted in the model, no studies investigated this area explicitly. Tuttle (1972) acknowledged the role of oral language in comprehension by hypothesizing that listening could be a more efficient learning mode for students who are blind, but he did not measure participants' oral language skills separate from his measures of listening comprehension. Nation (2005) described the interaction of reading comprehension and general language comprehension as a critical process, yet little is known about this interaction in children who read braille.

Within the framework of Perfetti et al.'s (2005) model of reading comprehension, some potentially important factors were not included. Creech and Golden (2009) used a token system to motivate their participant to read braille and answer comprehension questions correctly. Motivation, whether internal or external, is not reflected in Fig. 6.1, yet it has an impact on all the processes presented. If students are not motivated to read, they will not read. Another element missing from Fig. 6.1 is the availability of tactile input. Braille is not as prevalent as print, and children who are blind may not have opportunities to learn braille through incidental exposure. If there is nothing to read, then students cannot read, and they will not have sufficient prior knowledge to support the other components of the model.

4.5. Recommendations for Future Research

The primary focus of reading research in students who read braille has been on increasing reading rate and overall reading fluency. We found very little research regarding braille reading comprehension and its underlying processes. It is clear, when looking at Fig. 6.1, that research on braille reading has concentrated on tactile, perceptual issues. Furthermore, it is unclear whether current methods and measures for assessing reading comprehension are valid for students who read braille.

Most standardized tests do not include samples of students who are blind. In addition, most standardized tests are invalid if they are adapted in any way, including transcribing them into braille (Goodman, Evans, & Loftin, 2011). This problem extends to measures of variables that are highly correlated with reading ability, such as IQ and vocabulary. We must determine how to adequately measure reading comprehension in students who read braille, and these measures must have evidence of their validity.

The lack of basic information on braille reading comprehension is a real barrier to developing optimal interventions for students who read braille. There are factors we know to be associated with comprehension from typically developing students, such as vocabulary, and we should use these known factors as a starting place for braille reading research. Some researchers believe it is inappropriate to compare children who read braille to children with NV, because all findings could be attributable to the absence of vision, a variable that cannot be manipulated (Kelly & Smith, 2011). However, a comparison of this kind can be helpful if researchers are interested in comparing development of the two groups (Warren, 1994). Because little is known about the processes involved in reading braille, the comparative approach can provide valuable information on whether children who read braille are developmentally different from children with NV when it comes to reading comprehension. However, at the same time, we should be asking ourselves, "are the same factors associated with reading in braille?"

Many new technologies have emerged and are being widely used by braille readers, such as refreshable braille displays. One example is the prevalence of refreshable braille displays, which are portable and compatible with most iOS devices. However, refreshable braille displays can only display one line of braille at a time. What effect does this presentation of text have on comprehension? How does it compare to reading conventional braille? Some people might also question whether using a refreshable braille display is more efficient than using screen reading software.

There is reason to hypothesize that the auditory attention and memory capacity of students who are blind are larger than of sighted peers (Warren & Hatton, 2003). Students who are blind use auditory input as a learning medium more often than their peers, and this could make listening a more efficient alternative to reading at a certain point in their academic career. However, we have no research to support a shift from primarily reading braille to primarily listening other than the knowledge that reading braille is slow. It is important to know why listening might be more efficient and when a shift from reading to listening is needed. Additionally, reading only and listening only should be compared to audio-supported listening, a condition where reading and listening occur simultaneously.

It is possible that general knowledge deficits are the cause of gaps between reading ability and chronological age expectations (Gillon & Young, 2002). The age of onset of blindness, amount of functional vision, visual diagnosis, and the presence of additional disabilities are just some factors that could potentially affect the quality and quantity of early learning experiences in young readers of braille. Fewer experiences lead to less complete general background knowledge and vocabulary to draw upon when reading. We know vocabulary is highly correlated with reading ability in braille readers (Wall Emerson et al., 2009), yet there is no experimental research on vocabulary instruction. There is a need for research in this area, because children who are blind have little access to braille and fewer experiences than sighted peers, and it is within reason that systematic vocabulary instruction, and other strategies to compensate for the lack of experiential knowledge, can increase reading comprehension.

REFERENCES

Adams, M. J. (1990). *Beginning to read: Thinking and learning about print*. Cambridge, MA: MIT Press.

Brigance, A. H. (1999). *Brigance Comprehensive Inventory of Basic Skills-Revised (CIBS-R)*. North Billerica, MA: Curriculum Associates.

Campbell, A. F. (2011). Sarah's story: Using the individualized meaning-centered approach to braille literacy education (I-M-ABLE). *Journal of Visual Impairment & Blindness, 105*, 138–141.

Creech, J., & Golden, J. A. (2009). Increasing braille practice and reading comprehension in a student with visual impairment and moderate mental retardation: An initial study and follow-up. *Journal of Developmental and Physical Disabilities, 21*, 225–233. http://dx.doi.org/10.1007/s10882-009-9137-9.

Daneman, M. (1988). How reading braille is both like and unlike print. *Memory & Cognition, 16*, 497–504.

Daneman, M., & Newson, M. (1992). Assessing the importance of subvocalization during normal silent reading. *Reading and Writing, 4*, 55–77.

Davidson, P. W., Wiles-Kettenmann, M., Haber, R. N., & Appelle, S. (1980). Relationship between hand movements, reading competence and passage difficulty in braille reading. *Neuropsychologia, 18*, 629–635. http://dx.doi.org/10.1016/0028-3932(80)90102-5.

Dodd, B., & Conn, L. (2000). The effect of braille orthography on blind children's phonological awareness. *Journal of Research in Reading, 23*, 1–11.

Edmonds, C. J., & Pring, L. (2006). Generating inferences from written and spoken language: A comparison of children with visual impairment and children with sight. *British Journal of Developmental Psychology, 24*, 337–351.

Ferrell, K. A. (2006). Evidence-based practices for students with visual disabilities. *Communication Disorders Quarterly, 28*, 42–48.

Gillon, G. T., & Young, A. A. (2002). The phonological-awareness skills of children who are blind. *Journal of Visual Impairment & Blindness, 96*, 38–49.

Gilmore, J. V., & Gilmore, E. C. (1968). *Gilmore oral reading test*. New York: Harcourt, Brace & World.

Goodman, S. A., Evans, C., & Loftin, M. (2011). *Position paper: Intelligence testing of individuals who are blind or visually impaired*. Louisville, KY: American Printing House, Inc. http://www.aph.org/tests/intelligencetesting.htm.

Gough, P. B., & Tunmer, W. E. (1986). Decoding, reading, and reading disability. *Remedial and Special Education, 7*, 6–10.

Harley, R. K., & Rawls, R. (1970). Comparison of several approaches for teaching braille reading to blind children. *Education of the Visually Handicapped, 2*, 47–51.

Hong, S., & Erin, J. N. (2004). The impact of early exposure to uncontracted braille reading on students with visual impairments. *Journal of Visual Impairment & Blindness, 98*, 325–340.

Horner, R. H., Carr, E. G., Halle, J., McGee, G., Odom, S., & Wolery, M. (2005). The use of single-subject research to identify evidence-based practice in special education. *Exceptional Children, 71*, 165–179.

Johns, J. (2001). *Basic reading inventory: Pre-primer through grade twelve and early literacy assessments.* Dubuque, IA: Kendall/Hunt.

Kelly, S. M., & Smith, D. W. (2011). The impact of assistive technology on the educational performance of students with visual impairments: A synthesis of the research. *Journal of Visual Impairment & Blindness, 105*, 73–83.

Kintsch, W. (1998). *Comprehension: A paradigm for cognition.* Cambridge: Cambridge University Press.

Kintsch, W., & Rawson, K. A. (2005). Comprehension. In M. J. Snowling & C. Hulme (Eds.), *The science of reading: A handbook* (pp. 209–226). Oxford: Blackwell.

Kusajima, T. (1974). *Visual reading and braille reading: An experimental investigation of the physiology and psychology of visual and tactual reading.* New York: American Foundation for the Blind.

Lusk, K. E., & Corn, A. L. (2006). Learning and using print and braille: A study of dual-media learners, part 2. *Journal of Visual Impairment & Blindness, 100*, 653–665.

Martin, C. J., & Sheffield, C. (1976). The effect of telegraphic prose on the reading behavior of blind and sighted students. *Journal of Applied Psychology, 61*, 513–518. http://dx.doi.org/10.1037/0021-9010.61.4.513.

Millar, S. (1987). Perceptual and task factors in fluent braille. *Perception, 16*, 521–536.

Millar, S. (1990). Articulatory coding in prose reading: Evidence from braille on changes with skill. *British Journal of Psychology, 81*, 205–219.

Nation, K. (2005). Children's reading comprehension difficulties. In M. Snowling & C. Hulme (Eds.), *The science of reading: A handbook* (pp. 248–265). Malden, MA: Blackwell Publishing.

National Institute of Child Health, & Human Development (2000). Report of the National Reading PanelReports of the subgroups (NICHD Publication No. 00-4754). *Teaching children to read: An evidence-based assessment of the scientific research literature on reading and its implications for reading instruction.* Washington, DC: NICHD.

Nolan, C. Y., & Kederis, C. J. (1969). *Perceptual factors in braille word recognition.* New York: American Foundation for the Blind.

Perfetti, C. A., Landi, N., & Oakhill, J. (2005). The acquisition of reading comprehension skill. In M. Snowling & C. Hulme (Eds.), *The science of reading: A handbook* (pp. 227–247). Malden, MA: Blackwell Publishing.

Slowiaczek, M. L., & Clifton, C. (1980). Subvocalization and reading for meaning. *Journal of Verbal Learning and Verbal Behavior, 19*, 573–582.

Snow, C. (2002). *Reading for understanding: Toward an R&D program in reading comprehension.* Santa Monica, CA: RAND Corporation.

Steinman, B. A., LeJeune, B. J., & Kimbrough, B. T. (2006). Developmental stages of reading processes in children who are blind and sighted. *Journal of Visual Impairment & Blindness, 100*, 36–46.

Thompson, B., Diamond, K. E., McWilliam, R., Snyder, P., & Snyder, S. W. (2005). Evaluating the quality of evidence from correlational research for evidence-based practice. *Exceptional Children, 71*, 181–194.

Trent, S. D., & Truan, M. B. (1997). Speed, accuracy, and comprehension of adolescent braille readers in a specialized school. *Journal of Visual Impairment & Blindness, 91*, 494–500.

Tuttle, D. W. (1972). A comparison of three reading media for the blind: Braille, normal recording, and compressed speech. *Education of the Visually Handicapped, 4*, 40–44.

Wall Emerson, R., Holbrook, M. C., & D'Andrea, F. M. (2009). Acquisition of literacy skills by young children who are blind: Results from the ABC Braille Study. *Journal of Visual Impairment & Blindness, 103*, 610–624.

Warren, D. H. (1994). *Blindness and children: An individual differences approach*. Cambridge, England: Cambridge University Press.
Warren, D. H., & Hatton, D. D. (2003). Cognitive development of children with visual impairments. In I. Rapin & S. Segalowitz (Eds.), *Part II: Child neuropsychology: Vol. 8. Handbook of neuropsychology* (pp. 439–458) (2nd ed). New York: Elsevier.
Wormsley, D. P. (2011). A theoretical rationale for using the individualized meaning-centered approach to braille literacy education with students who have mild to moderate cognitive disabilities. *Journal of Visual Impairment & Blindness, 105*, 145–156.
Wright, T. (2010). Looking for power: The difficulties and possibilities of finding participants for braille research. *Journal of Visual Impairment & Blindness, 104*, 775–780.
Wright, T., Wormsley, D. P., & Kamei-Hannan, C. (2009). Hand movements and braille reading efficiency: Data from the Alphabetic Braille and Contracted Braille Study. *Journal of Visual Impairment & Blindness, 103*, 649–661.

CHAPTER SEVEN

Three Areas of the Expanded Core Curriculum for Students with Visual Impairment: Research Priorities for Independent Living Skills, Self-Determination, and Social Interaction Skills

Sandra Lewis[*,1], **Mackenzie E. Savaiano**[†], **Karen Blankenship**[†], **Kitty Greeley-Bennett**[*]

[*]School of Teacher Education, Florida State University, Tallahassee, Florida, United States
[†]Department of Special Education, Vanderbilt University, Nashville, Tennessee, USA
[1]Corresponding author: e-mail address: slewis@fsu.edu

Contents

Abstract

An extensive review of existing literature was conducted to identify intervention research in three areas of the expanded core curriculum (ECC) for students with visual

International Review of Research in Developmental Disabilities, Volume 46
ISSN 2211-6095
http://dx.doi.org/10.1016/B978-0-12-420039-5.00002-2

impairments: independent living skills, self-determination skills, and social interaction skills. Students with visual impairments often struggle to develop competence in these three areas without direct instruction, yet it is thought that they are critical for successful adult functioning. After identifying over 700 potential articles through an extensive literature search, 23 intervention studies on school-age children with visual impairments met inclusion criteria for this chapter. Results are summarized by the ECC area, with seven (30%) studies on independent living skills, six (26%) studies on self-determination, and ten (43%) studies on social skills. Single-subject multiple baseline designs were most commonly used, though many of the researchers failed to demonstrate experimental control through sufficient replications of their interventions to support causality. Based on the paucity of intervention studies, the authors conclude that there are few research-based strategies to facilitate the development of skills in these three areas. More research, and research of higher quality, is needed.

1. INTRODUCTION

First framed by Hatlen (1996), the expanded core curriculum (ECC) identified the academic and functional areas most impacted by the presence of a visual impairment (blindness or low vision) and in which direct instruction is likely to be necessary. The ECC originally included eight content areas: assistive technology/technology, career education, compensatory or functional academics, independent living skills, orientation and mobility, recreation and leisure, visual efficiency, and social interaction skills (Sapp & Hatlen, 2010). These content areas have evolved to include additional areas critical for the successful achievement of long-term outcomes by students with visual impairments. For instance, in 2003, self-determination was added and the area originally known as compensatory or functional academics was changed to compensatory access skills, reflecting the increased focus of the *Individuals with Disabilities Educational Improvement Act* (IDEA, 2004) on the general curriculum for all children. Similarly, in 2009, the area once referred to as visual efficiency was expanded to sensory efficiency skills (visual, auditory, and tactual) upon recommendations from the field to recognize learning and literacy skills needed by all children with blindness or low vision, including children with more severe disabilities (Phil Hatlen, personal communication, 2009). Table 7.1 provides definitions for the nine areas of the ECC.

The ECC is designed to provide access to the core curriculum components of math, reading, writing, and science and to address essential areas and experiences unique to persons who have visual impairments (Pugh & Erin,

Table 7.1 Areas of the Expanded Core Curriculum

ECC Area	Definition
Assistive technology	Selection of and instruction in the use of assistive technology devices to improve functional outcomes of children with visual impairments
Career education	Provide access to and understanding of careers of which children with visual impairment may not be aware. This area also addresses the vocational skills students need to perform jobs and maintain employment
Compensatory access	Skills that students with visual impairment need to successfully access all areas of the core curriculum. Mastery of compensatory skills usually means that students have access to learning in a manner equal to that of sighted peers
Independent living	All the tasks and functions people perform, according to their abilities, in order to live as independently as possible
Orientation and mobility	Skills that help people who are visually impaired to travel as independently as possible; includes body awareness and travel in rural and urban environments
Recreation and leisure	Provide access to the choices available and provide instruction in how to perform leisure skills that most students learn through observation. Includes traditional and adapted physical education activities and games and hobbies
Self-determination	A combination of skills, knowledge, and beliefs that enable a person to engage in goal-directed, self-regulated, autonomous behavior composed of self-awareness, self-advocacy, choice making, self-management/self-regulation, decision making, self-efficacy, and goal setting and attainment
Sensory efficiency	Instruction in the use of residual vision, hearing, and other senses in order to access information for learning and literacy
Social interaction	Development of skills related to interpreting and using body language and other nonverbal communication, appropriate social communication strategies, conversation patterns, and interactions; developing friendships and relationships; and monitoring one's social behavior

1999). These areas should be taught in addition to the core curriculum because they are specific to the disability of blindness or require instructional methods that may differ considerably from those used by teachers of children with unimpaired vision.

Children with visual impairments often have limited, if any, opportunities to learn the skills covered within the ECC, except through sequential, systematic instruction. Instruction in these skills requires understanding the impact of visual impairment on learning and competency in teaching these skills using nonvisual strategies. Teachers of students with visual impairments (TVIs) and orientation and mobility specialists have the requisite knowledge to teach these students. Blankenship (2007; 2008) posited that the ECC is the content knowledge of TVIs and orientation and mobility specialists and should be a focus of direct instruction for all students with visual impairment as required by IDEA, 2004. Interestingly, Wolffe et al. (2002) found that most TVIs report that they spend the majority of their time tutoring or providing access to general education, not providing specially designed instruction in the ECC content areas.

It has taken the field of visual impairments many years to achieve a collaborative understanding of the framework and terminology to advocate for instruction in the ECC. A number of research studies report an increase in the awareness, agreed upon importance, and implementation of the ECC since 1996 (Brown & Beamish, 2012; Hatlen & Blankenship, 2010; Lohmeier, Blankenship, & Hatlen, 2009; Sapp & Hatlen, 2007; Wolffe, Hatlen, & Blankenship, 2011). McKenzie and Lewis (2008) reported that 37% of TVIs surveyed admitted to assigning paraprofessionals direct instructional tasks in at least one area of the ECC, even though these paraprofessionals were not certified to provide such instruction.

It is not within the scope of one chapter to review the research in all nine areas of the ECC. It would also be redundant with other chapters to cover assistive technology (see Chapter 3) or orientation and mobility (see Chapter 8). The purpose of this chapter was to describe the intervention research and identify research priorities in three areas of the ECC: (a) independent living, (b) self-determination, and (c) social interaction skills.

1.1. Independent Living Skills

The term independent living skills is a broad label that encompasses every skill that an individual needs to have at some point in his or her life to be as independent as possible (Hatlen, 1996). Over the years, these activities have been referred to as activities of daily living (ADL), daily living skills, or adapted living skills. The skills associated with independent living, especially basic hygiene and dressing, are believed to be associated with successful career transition (Crudden, 2012). Although the term is very broad, for this chapter, independent living skills relate to the constructs of (a) personal management (e.g., grooming, hygiene, and dressing), (b) food management (e.g., eating, cooking, and shopping), (c) home management (e.g., cleaning), and (d) money management (e.g., budgeting, obtaining and using money, and balancing a checkbook).

Sacks, Wolffe, and Tierney (1998) interviewed 48 students and their parents. One component of the interview covered independent living skills, and the authors found that students with visual impairment were similar to sighted peers in money, time, and personal management, although there was a pattern suggesting that the level of independence was associated with the level of functional vision. Students with visual impairments reported much lower participation in home management than sighted peers. Another survey of parents found significant differences in the perceived independent living skills of children with visual impairments and their sighted peers, aged 6–9 years (Lewis & Iselin, 2002). Parents of sighted children indicated that their children performed 84% of the identified tasks without assistance and 4% of the tasks with assistance, while parents of children with visual impairments reported that their children performed only 44% of the tasks independently and 14% the tasks with assistance. These findings suggest that the performance of independent living skills is different for children with visual impairments than for same-age sighted peers.

This disparity, for some reason, has been identified most often in research on the oral hygiene of students with visual impairments (Ameer et al., 2012; Anaise, 1979; Greeley, Goldstein, & Forrester, 1976; Ozdemir-Ozenen, Sungurtekin, Cildir, & Sandalli, 2012). We can be reasonably certain that children who are blind have more plaque and decay than children with normal sight (Ameer et al., 2012; Anaise, 1979; Ozdemir-Ozenen et al., 2012) or children who are "partially blind" (Anaise, 1979; Greeley et al., 1976).

1.2. Self-Determination

Self-determination has been defined as a combination of skills, knowledge, and beliefs that enable a person to engage in goal-directed, self-regulated, autonomous behavior composed of self-awareness, self-advocacy, choice making, self-management/self-regulation, decision making, self-efficacy, and goal setting and attainment (Field, Martin, Miller, Ward, & Wehmeyer, 1998; Lane, Carter, & Sisco, 2012; Self-Determination Technical Assistance Center, 2000). An understanding of one's strengths and limitations and a belief in oneself as capable and effective are essential to self-determination. When acting on the basis of these skills and attitudes, individuals have a greater ability to take control of their lives and assume the role of successful adults.

Agran, Hong, and Blankenship (2007) found that TVIs reported being familiar with the term self-determination, thinking it represented important skills needed by their students. In addition, 27% of the respondents noted that none of their students had self-determination goals on their IEPs, while 10% stated that all of their students had such goals. The reasons given for not providing direct instruction in self-determination were the lack of curricula, lack of familiarity with the construct, insufficient time, and more pressing curricular needs for their caseload. Lewis and McKenzie (2010) found that TVIs were providing instruction in self-determination to students enrolled both in special purpose (11.1%) and in local schools (25.7%) and that this instruction was also being provided by paraprofessionals in these locations (37% at special purpose and 13.7% at local schools). In another paper using results from the same survey, McKenzie and Lewis (2008) found that paraprofessionals reported that they had never received training in the provision of instruction in the area of self-determination.

A majority of research on self-determination has focused on the construct of self-awareness. Previous research has shown that students with visual impairment scored either the same or higher than sighted peers on measures of self-esteem and self-concept (Griffin-Shirley & Nes, 2005; Obiakor & Stile, 1989).

1.3. Social Interaction Skills

The social interaction skills area of the ECC includes the development of skills related to interpreting and using (a) body language and other nonverbal communication; (b) appropriate social communication strategies, conversation patterns, and interactions; (c) developing friendships and relationships; and (d) monitoring one's social behavior. Instruction in object and social play, turn taking, understanding one's feelings and the feelings of others, social concepts, assertiveness skills, and a myriad of other behaviors are often necessary to achieve successful socialization. Little is known, however, about efficacious strategies for teaching these skills.

Children with visual impairments face many challenges in the development of behaviors that facilitate socialization. The natural process of the development of these skills in unimpaired children involves observations to accurately interpret social environments, to appreciate the contexts in which socialization occurs, to take the perspective of others, and to shape social behaviors. Without accurate information about the complexity of engaging in and maintaining a variety of social interactions, children who have visual impairments frequently are at a social disadvantage. Instruction in social skills, then, is thought to be necessary to overcome this disadvantage (Sacks & Silberman, 2000).

The purpose of this chapter is to identify strategies for facilitating the development of (a) independent living skills, (b) self-determination skills, and (c) social interaction skills in students with visual impairments by reviewing intervention research. There is significant overlap of these functional nonacademic skills, all of which are essential to healthy postschool personal and career success.

2. METHOD

Searches were conducted using ERIC, PsycINFO, and ProQuest Central databases for each area of the ECC covered in this chapter. No delimiters for the date of publication were entered in the search. General search procedures also included hand searches of the following periodicals: the *British Journal of Visual Impairment*, *Education of the Visually Handicapped*, *Exceptional Children*, *Insight*, *International Journal of Special Education*, *Journal of Applied Behavior Analysis*, *Journal of Visual Impairment & Blindness*, *New Outlook for the Blind*, *Outlook for the Blind*, *RE:view*, *TEACHING Exceptional Children*, and *Topics in Early Childhood Special Education*. Once articles were identified in each area as meeting inclusion criteria, ancestral searches were conducted to make sure no relevant research articles were missed.

2.1. Inclusion and Exclusion Criteria

To be included in this chapter, articles had to (a) include at least one participant with a visual impairment, (b) include data collection on the ECC area of interest, (c) include school-age children aged 3–22 years, (d) be peer-reviewed, and (e) include research on an intervention. Articles were excluded if (a) they were case studies, (b) they were primarily about orientation and mobility, or (c) the results of school-age participants could not be disaggregated from results of a larger sample including adult participants.

2.2. Search Terms

All searches included the terms "visual handicap" or "vis* impair*" or "low vision" or "partial sight" or "partial vision" or "blind*" and "child" or "youth" and "instruction" or "intervention." Specific search terms related to independent living skills included "independent living" or "daily living" or "functional skills" or "adaptive skills" or "unique skills" or "disability specific skills" or "self care" or "fundamental skills." Specific search terms related to self-determination included "self determination" or "problem solving" or "self management" or "decision making" or "goal setting" or "choice making" or "self concept" or "self efficacy" or "self empowerment" or "self motivation" or "self esteem" or "personal values" or "self regulation" or "self advocacy" or "locus of control." Specific search terms related to social interaction included "social interaction" or "social skills" or "interpersonal competence" or "relationship" or "friendship" or "dating" or "parallel play" or "group play" or "symbolic play" or "pretend play" or "social concepts" or "assertiveness training" or "turn taking."

3. RESULTS

3.1. Independent Living Skills

Database searches returned 342 citations, and hand searches returned an additional 87 citations. Of these articles, 50 (11%) were eliminated for having adult participants or results that could not be disaggregated from the larger sample, 237 (55%) were eliminated because data were not collected on a construct or specific skill related to independent living skills, 47 (11%) were eliminated because participants did not have visual impairments, and 95 (22%) were eliminated by other exclusion criteria (e.g., not intervention research). A total of seven (1%) intervention studies met the criteria for inclusion.

Table 7.2 describes the components of the seven independent living skills intervention studies identified in this chapter. The constructs covered include food management (i.e., eating, drinking, food preparation, and grocery shopping; $n=5$), oral hygiene ($n=1$), and money ($n=1$). Studies were conducted in Israel, Taiwan, and the United States; participants' ages ranged from 6 to 21 years; and articles were published between 1969 and 2012. All seven studies used single-subject designs.

Table 7.2 Summary of ILS Intervention Studies Included in this Chapter

	N	Age of Participant(s) (in Years)	Visual Information	Specific Skill or Construct	Design	Type of Intervention	Results	Functional Relation	Geographic Location
Farrenkopf, McGregor, Nes, and Koenig (1997)	1	17	CVI	Drinking	SSD-ATD combined with MB across 2 settings	1 = Verbal prompt: students' name, stimulus object, and direction (e.g., "[Student's name] take a drink from your cup") 2 = Physical prompt: place hand, palm down, on top of student's head	Physical prompts were more effective than verbal prompts in increasing independent drinking skills. The physical prompting intervention was also introduced in a time-lagged fashion in the home and produced similar results	Yes	Texas

Continued

Table 7.2 Summary of ILS Intervention Studies Included in this Chapter—cont'd

	N	Age of Participant(s) (in Years)	Visual Information	Specific Skill or Construct	Design	Type of Intervention	Results	Functional Relation	Geographic Location
Gumpel and Nativ-Ari-Am (2001)	4	P1 = 19 P2 = 21 P3 = 17 P4 = 17	P1 = Lawrence syndrome, 1/60 P2 = Behr's syndrome, 1/60 P3 = cataracts, nystagmus, 6/60 P4 = 6/60	Grocery shopping	SSD-MB across dyads	1 = Task analysis: instruction provided for individual steps of grocery shopping (17 chained skills) 2 = Self-monitoring: students received a set of 10 task cards to use during grocery shopping	Task analysis immediately increased students' ability to purchase correct items, but not on correct sequence of skills; Self-monitoring stabilized, but did not increase students' ability to purchase correct items or complete the correct sequence of skills	No	Israel

Hanney and Tiger (2012)	2	PI=6 P2=8	P1=OCA, nystagmus, photophobia P2=retinal detachment, cataract (OS), NLP (OD)	Money	SSD-MB across 2 behaviors	Instructed students to select a coin given the coin's name (name–coin) and to select a coin given the coin's value (value–coin)	Instruction increased correct responses for both participants and instruction generalized to other associations between coins and their values (i.e., coin–name, name–value, coin–value, and value–name)	No	School for the blind in the United States (state not reported)
Shih and Chang (2005)	3	P1=8 P2=8 P3=7	P1=low vision P2=blind P3=blind	Oral hygiene	SSD-ABB^{1}A^1	Oral hygiene program: Part 1=classroom discussions (with models) about structure and functions of the mouth,	All three participants' plaque decreased from baseline (A) to intervention (B)	No	Taipei, Taiwan

Continued

Table 7.2 Summary of ILS Intervention Studies Included in this Chapter—cont'd

	N	Age of Participant(s) (in Years)	Visual Information	Specific Skill or Construct	Design	Type of Intervention	Results	Functional Relation	Geographic Location
						benefits of brushing, and using dental floss Part 2 = individual coaching from a dentist			
Sisson and Dixon (1986)	1	10	Blind	Eating	SSD-ABAB for posture, combined with MB across 2 behaviors	Posture: story on tape contingent on good posture (hands down, head up) Eating behaviors: praise and reinforcement for correct behavior. Inappropriate behavior resulted in "time-out"	Contingent story increased appropriate posture Praise and "time-out" appear to have increased napkin use and utensil use	Posture: yes Eating: no	Penn.

Stolz and Wolf (1969)	1	16	Optic nerve atrophy	Eating	SSD-ABA^{1}B for cafeteria line	Cafeteria line: researchers told aides to stop helping in the cafeteria line, and then, this help was reinstated and removed again	Cafeteria line: no immediate change from baseline (A) to intervention (B), but an overall change in trend. Also, distinct change in level from (B) to (C) and back to (B)	Cafeteria line: yes	Kansas
					SSD-AB for eating	Eating: Participant was taught the use of a fork, correct utensils to use for different kinds of food, and "other aspects" of appropriate eating	Eating: Overall change in level from baseline (A) to intervention (B), but there wereas a lot of overlapping data.	Eating: no	Kansas

Continued

Table 7.2 Summary of ILS Intervention Studies Included in this Chapter—cont'd

	N	Age of Participant(s) (in Years)	Visual Information	Specific Skill or Construct	Design	Type of Intervention	Results	Functional Relation	Geographic Location
Taylor (1987)	1	21	Deafblind	Food preparation	SSD-MP across 3 behaviors	Participant was provided with tactile recipe cards (miniature objects glued to the card). In baseline, no instruction was given. During intervention, the trainer guided the participants' hand through the step, then put the item back, and gave the prompt for her to do it independently	Prompt training was effective for teaching preparation of all three foods (cereal, chocolate milk, and juice). Although the intervention appears effective in all tiers, there was no probe data immediately before the introduction of the intervention in tier 2 or tier 3	No	Kansas

CVI, cortical visual impairment; SSD, single-subject design; ATD, alternating treatments design; MB, multiple baseline; OCA, oculocutaneous albinism; OS, left eye; NLP, no light perception; OD, right eye; MP, multiple probe.

3.1.1 Food Management

Three of the five research articles on food management had some focus on appropriate eating or drinking skills (Farrenkopf et al., 1997; Sisson & Dixon, 1986; Stolz & Wolf, 1969), one examined food preparation (Taylor, 1987), and one examined grocery shopping (Gumpel & Nativ-Ari-Am, 2001). Three of these articles (Farrenkopf et al., 1997; Sisson & Dixon, 1986; Stolz & Wolf, 1969) demonstrated a functional relation between their intervention and a change in behavior, but only one research group (Farrenkopf et al., 1997) demonstrated an effect for the actual eating behavior, finding that physical prompts were more effective than verbal prompts in increasing a student's drinking skills.

3.1.2 Personal Management (Oral Hygiene)

Shih and Chang (2005) taught three students with visual impairment about dental hygiene to see if intervention decreased students' plaque levels. A teacher of students with visual impairment led classroom discussions, and a dentist conducted individual coaching sessions. In addition to standard oral hygiene skills, the dentist taught students to feel for plaque on the surfaces of their teeth with their fingers and tongues. All three participants had less plaque after intervention, but there were not enough replications to demonstrate a functional relation.

3.1.3 Money Management

Hanney and Tiger (2012) trained two students with visual impairment to select a coin, based on the size of the coin, given the coin's name, and to select a coin given the coin's value. Training only occurred for those two behaviors, but Hanney and Tiger (2012) collected data on four other behaviors: (a) say the coin's name given the coin, (b) say the value of the coin given its name, (c) say the value of the coin given the coin, and (d) say the coin's name given its value. Training appeared to increase both students' correct responses, and training on two behaviors appeared to increase correct responses on all four untrained behaviors (generalization), but there were not enough replications to show a functional relation between the intervention and the change in behavior.

3.2. Self-Determination

Database searches returned 629 citations, and hand searches returned an additional 88 citations. Of these articles, 34 (5%) were eliminated for having adult participants or results that could not be disaggregated from the larger sample, 270 (38%) were eliminated because data were not collected on a construct or specific skill related to self-determination, 287 (40%) were eliminated because participants did not have visual impairments, and 120 (16%) were eliminated by other exclusion criteria (e.g., not intervention research). A total of six (1%) intervention studies met the criteria for inclusion.

Table 7.3 describes the components of the six self-determination intervention studies identified in this chapter. The constructs covered included self-awareness ($n=3$), goal setting ($n=1$), self-management ($n=1$), and all constructs/components ($n=1$). Studies were conducted in Nigeria, Egypt, India, and the United States; participants' ages ranged from 8 to 21 years; and articles were published between 1996 and 2011. Some researchers did not provide information about the age and visual status of their participants; however, all of the researchers except Jindal-Snape (2005b), who included one 9-year-old who was totally blind, appeared to include students with low vision and blindness.

3.2.1 Self-Awareness

Two of three research groups whose goal was to improve self-awareness reported that their intervention was effective (Mohamed, Bayoumi, & Diaz 2011; Shapiro, Moffett, Lieberman, & Dummer 2005). Mohamed et al. (2011) conducted a medical intervention study designed to improve self-awareness and management of glaucoma in 50 12- to 18-year-olds in Egypt and found a significant effect of the treatment on participants. Although Shapiro et al. (2005) reported a main effect of improved perceptions of social acceptance, athletic competence, and physical appearance for all 43 students who attended a week-long athletic camp, descriptive statistics suggest that females were the only students who showed improvement in these three areas of self-awareness. Alexander (1996) reported that his weekly group intervention with 20 third to fifth graders designed to improve self-concept/self-awareness was not effective.

3.2.2 Goal Setting

Eniola and Adebiyi (2007) described a pre-/poststudy to improve motivation to work in students with visual impairments in Nigeria. Two interventions were used with the two groups of 16 students. The authors did not provide the ages or visual status of the participants. One intervention was described as focusing on emotional intelligence and the other as focusing on goal setting. Although the authors noted that the intervention was provided twice weekly for 6 weeks, it is unknown for how long each session lasted.

Table 7.3 Summary of Self-determination Intervention Studies Included in this Chapter

	N	Age of Participant(s) (in Years)	Visual Information	Specific Skill(s) or Construct(s)	Design	Type of Intervention	Results	Geographic Location
Alexander (1996)	20	18–11	Not reported (all were legally blind)	Self-awareness	Pre-/posttest	Components from *Esteem Builders curriculum* (Borba, 1991) were implemented for 18 weeks (once weekly sessions, unknown length)	No significant differences were found between day versus residential students and braille versus print readers. No main effect of treatment	Tennessee
Eniola and Adebiyi (2007)	32	Not reported	Not reported	Goal setting	Pre-/posttest	2 Groups of 16 received either emotional intelligence or goal-setting interventions to increase motivation to work for 6 weeks (twice weekly sessions of lecture, demonstration, discussions with homework, and activity assignments)	Positive effect overall but not significant effects by group (emotional intelligence/goal setting) or gender Authors suggested that descriptive means might favor emotional intelligence intervention	Nigeria

Continued

Table 7.3 Summary of Self-determination Intervention Studies Included in this Chapter—cont'd

	N	Age of Participant(s) (in Years)	Visual Information	Specific Skill(s) or Construct(s)	Design	Type of Intervention	Results	Geographic Location
Jindal-Snape (2005b)	1	9	Not reported	Self-management	SSD-MB across 2 behaviors	Researcher prompted student to self-evaluate for direction of gaze and on-task behavior after 3-min intervals (two 10-min sessions per day). A peer provided feedback after self-evaluations of direction of eye gaze. Intervention lasted 5 weeks	Direction of gaze and on-task behaviors increased but not enough replications to demonstrate a functional relation Author suggested that peer feedback enhanced self-evaluation	India
Levin and Rotheram-Fuller (2011)	30	14–21	Not reported	All	Quasi-experimental pre-/post	2 Cohorts of 15; one cohort received intervention using the empowered curriculum (Cleveland et al.,	No effects of intervention were found	Northeastern United States

						2007) during two 45-min sessions weekly for 15 weeks. Cohorts were not randomly assigned and both received pre-/postmeasures		
Mohamed, Bayoumi, and Draz (2011)	50	12–18	Glaucoma	Self-awareness	Pre-/posttest	Participant dyads were trained on glaucoma management (administration of eye drops, ointment, compresses, and exercises) for 15 sessions in context of medical clinic	Significant effect of treatment on knowledge of glaucoma; physical, social, and psychological questionnaires; expectations; and skills needed to manage glaucoma	Egypt
Shapiro, Moffett, Lieberman, and Dummer (2005)	43	8–21	Not reported (USABA categories provided)	Self-awareness	Pre-/posttest	Student participants in 1-week summer camp completed three pre-/postmeasures to examine whether	Main effect for camp reported; perceived self-awareness was higher on posttests. Significant	The United States

Continued

Table 7.3 Summary of Self-determination Intervention Studies Included in this Chapter—cont'd

	N	Age of Participant(s) (in Years)	Visual Information	Specific Skill(s) or Construct(s)	Design	Type of Intervention	Results	Geographic Location
						summer camp improved the perception of social acceptance, physical appearance, or athletic competence	interaction between gender and camp attendance—boys' perceptions higher at pretest, but no difference at posttest. Descriptive statistics suggested that self-awareness of girls, but not boys, improved during camp	

SSD, single-subject design; MB, multiple baseline; USABA, United States Association of Blind Athletes.

Sessions consisted of lectures, demonstrations, and discussions, and the authors noted that participants also had homework and activities to complete in addition to time in the sessions. The authors reported that the intervention worked and that there were no group or gender interactions; however, they noted that group means suggested that the emotional intelligence intervention might be more effective.

3.2.3 Self-Management

Results from Jindal-Snape's (2005b) single-subject research study suggest that self-evaluation was used effectively by a 9-year-old with total blindness in an inclusive education classroom in New Delhi; however, the author did not demonstrate a functional relation that would document that improved direction of eye gaze and on-task behavior were related to the student's ability to self-evaluate. The peer feedback component of the study seemed promising, but it was used only for the direction of eye gaze behavior, not time on-task. Although the author suggested that peer feedback enhanced self-evaluation, visual inspection of the data did not support that assertion.

3.2.4 All Constructs of Self-Determination

Levin and Rotheram-Fuller (2011) used the empowered curriculum (Cleveland et al., 2007) and a pre-/postdesign with two cohorts of 15 students, aged 14–21 years. All students were assessed pre- and posttreatment, but only one cohort received the intervention. Twice weekly sessions of 45 min were held for 15 weeks, following instructions for all 23 units of the empowered curriculum. No significant differences between the two groups were reported on measures of self-determination, self-concept, and self-perception of emotions.

3.3. Social Interaction Skills

Database searches returned 543 citations, and hand searches returned an additional 151 citations. Of these articles, 31 (4%) were eliminated for having adult participants or results that could not be disaggregated from the larger sample, 184 (26%) were eliminated because data were not collected on a construct or specific skill related to social skills, 216 (31%) were eliminated because participants did not have visual impairments, and 266 (38%) were eliminated by other exclusion criteria (e.g., not intervention research). A total of 10 (1.4%) intervention studies met the criteria for inclusion.

Table 7.4 describes components of the 10 social skills intervention studies identified in this chapter. The main constructs covered include conversation

Table 7.4 Summary of Social Skills Intervention Studies Included in this Chapter

	N	Age of Participant(s) (in Years)	Visual Information	Specific Skill(s) or Construct(s)	Design	Type of Intervention	Results	Geographic Location
D'Allura (2002)	13	46–52 months	Achromatopsia Albinism Cataracts Cortical blindness Juvenile MD ROP	Interactions (with peers)	Static-group comparison design	Trained inclusive classroom teacher to implement cooperative learning strategies, compared to self-contained classroom that was not trained	Children with VI in inclusive classroom interacted with peers at levels similar to peers and were more likely to initiate with peers; social interactions of children with VI in self-contained classroom remained the same	New York City
Gronna, Serna, Kennedy, and Prater (1999)	1	30 months	Severe esotropia	Interactions (with others), conversation patterns	SSD-MB across 5 skills	Puppets were used to train participant plus 4 peers using scripts and discussion. Training sessions were followed by free-play sessions	Participant was described as "socially isolated" prior to the intervention.	Hawaii

							During intervention, target behaviors approached levels of 100% and teacher rating increased from 1.8 to 5	
Jindal-Snape (2004)	2	P1: 10 P2: 9.10	P1: NLP P2: LP	Body language (eye gaze), conversation patterns	SSD–combination ABACA(BA)[a] and MB across 2 behaviors	Self-evaluation and feedback were used to check behaviors every 2 min during conversations with peers	Results are reported separately for each subject, but both showed increases in duration and number of social interactions with peers from levels near 0% to levels near 100%	India

Continued

Table 7.4 Summary of Social Skills Intervention Studies Included in this Chapter—cont'd

	N	Age of Participant(s) (in Years)	Visual Information	Specific Skill(s) or Construct(s)	Design	Type of Intervention	Results	Geographic Location
Jindal-Snape (2005a)	1	9.4	Blind	Body language (eye contact), conversation patterns	SSD–combination ABCBA and MB across 2 behaviors	Participant was trained to recruit feedback from peers and in turn trained peers to give feedback on social behaviors. Self-evaluation and self-evaluation + feedback phases were used	Eye contact (gaze) and sustained conversation skills improved. Self-recruitment of feedback increased the accuracy of self-evaluation and peers extended their feedback to other social environments	India
Kim (2003)	26	13–19	ROP ONH Aniridia Myopia Optic atrophy Others	Body language, conversation patterns	2 Group experiments, pre-/posttest	Implemented Assertiveness Training Curriculum for Adolescents with VI (ATCAVI) using instruction and role-play. Students	No significant treatment effect; both groups showed improvement. Difference reported in	Southeastern United States

						randomly assigned to treatment or control group. Parents and a teacher filled out questionnaires for pre-/postmeasures	parent-rated scores, almost significant difference in student-reported ratings, and no difference in observation scores	
Mar and Sall (1995)	3	P1: 10 P2: 10 P3: 7	P1: ONH P2: cortical blindness P3: retinal vascular aplasia All dual-sensory loss, 1 in general education, 2 with severe cognitive impairments	Conversation patterns, social communication, development of relationships (social networks)	Pre-/posttest	Interventions were individualized for each student and consisted of team meetings, educator support, or in-service trainings; modeling; or parent support every 2 weeks during a school year. Interviews, checklists, and a social network analysis were conducted pre- and postintervention	Increases were found in social opportunities, in social activities in the community, and in the social networks of all three students, though the amount of increase varied for each participant	New York City

Continued

Table 7.4 Summary of Social Skills Intervention Studies Included in this Chapter—cont'd

	N	Age of Participant(s) (in Years)	Visual Information	Specific Skill(s) or Construct(s)	Design	Type of Intervention	Results	Geographic Location
Raver (1984)	1	3.5	ONH	Body language (head droop)	SSD-ABAB reversal design	Direct instruction from teacher and paraprofessional followed by "Look At Me" game. Prompted and spontaneous head turns were heavily praised and rewarded	High increases in performance during intervention, with maintenance and generalization levels higher than baseline, but lower than intervention	Not reported
Sacks and Gaylord-Ross (1989)	20	7.2–12.2	15 w/VI: glaucoma ONH ROP Albinism Retinoblastoma Optic nerve disorders Microphthalmia (All LB)	Body language (eye gaze and posture), conversation patterns (groups)	SSD-MB across 7 behaviors	Facilitation of social skills instruction using peer-mediated and teacher-directed approaches. Peers trained through weekly sessions with researcher, who also conducted training of subjects for teacher-directed approach	All participants in experimental groups increased demonstration of targeted behaviors. Only students trained with peer-mediated	California

							approach demonstrated skills in natural settings during the follow-up phase of study	
Sisson, Van Hasselt, Hersen, and Strain (1985)	6	P1: 11.7 P2: 10.9 P3: 10.5 P4: 9.3 2 peers	P1: aniridia Cataracts Macular hypoplasia P2: cortical blindness P3: myopia Elliptical optic nerves P4: cataracts Aphakia (All LB)	Conversation patterns, interactions (with peers)	SSD-MB 2 MB studies across 2 participants (boys and girls treated as separate studies)	Peers were trained through simulations and instruction on how to initiate play and interactions with children with multiple disabilities. Peers then joined free play in the other students' classroom and received "booster" training sessions during 6 weeks of intervention	All participants, including peers, increased their social behaviors. Follow-up 4 months after the intervention showed levels higher than intervention	Pittsburgh, PA
Yarnall (1979)	1	6.5	Severe hearing loss, cataract extractions	Body language (eye gaze)	SSD; AB	Intervention involved using food reinforcement held beside the teacher's	Child's eye contact increased from nearly 0 to	School for the blind in the United States

Continued

Table 7.4 Summary of Social Skills Intervention Studies Included in this Chapter—cont'd

	N	Age of Participant(s) (in Years)	Visual Information	Specific Skill(s) or Construct(s)	Design	Type of Intervention	Results	Geographic Location
						eye to elicit eye contact. Child received food and social reward. Use of the food was faded as child maintained eye contact	more than 300 s during a 600-s session on 5 consecutive sessions. Generalization to other environments was reported, but not documented	

[a]One participant had an ABACA design and one participant had an ABACABA design. Both participants had SSD-MB across 2 behaviors.
MD, macular degeneration; ROP, retinopathy of prematurity; VI, visual impairment; SSD, single-subject design; MB, multiple baseline; NLP, no light perception; LP, light perception; ONH, optic nerve hypoplasia; LB, legally blind.

patterns ($n=7$), body language and gestures ($n=6$), interactions with others ($n=3$), development of relationships and friendships ($n=1$), and social communication ($n=1$). Seven articles covered more than one construct. Studies were conducted in India and the United States, participants' ages ranged from 30 months to 19 years; and articles were published between 1979 and 2005. Seven (70%) studies used a single-subject design, two (20%) used a pre-/posttest to measure change in one or two groups, and one (10%) study was categorized as using a static-group comparison design.

3.3.1 Preschoolers

Three of the interventions that were reviewed included subjects who were of preschool age. Two of these studies used a single-subject design and one used a static-group comparison design; all three interventions showed an increase in the targeted social interaction construct. Two studies (Gronna et al., 1999; Raver, 1984) reported effective interventions that were designed to increase the social interactions of young preschool children. D'Allura (2002) found significant differences in the length and frequency of social interactions in an inclusive preschool classroom implementing cooperative learning strategies compared to a self-contained classroom that was not using cooperative learning strategies.

3.3.2 Children in Middle Childhood

Four of the interventions targeted children who were in early or middle childhood (ages 9–12 years), and all used a single-subject design. Of the four studies in middle childhood, three (Jindal-Snape, 2005a; Sacks & Gaylord-Ross, 1989; Sisson et al., 1985) found that using peer-mediated interventions increased the frequency and generalizability of targeted social behaviors in children with visual impairments. Sacks and Gaylord-Ross (1989) found no significant differences between adult-directed and peer-mediated interventions, though there was higher generalization of social skills from the peer-mediated group. Jindal-Snape (2005a) and Sisson et al. (1985) also found that trained peers generalized their social skills facilitation to other environments. Jindal-Snape (2004, 2005a) found that self-evaluation can increase targeted social skills when adult or peer feedback is provided. However, all studies except Sacks and Gaylord-Ross (1989) failed to demonstrate functional relations between their intervention and the increase in social skills.

3.3.3 *Students with Deafblindness*

Two intervention studies addressed the social communication skills of children who were deafblind who experienced particularly onerous challenges to communication and socialization. Yarnall (1979) used a single-subject design and reported that operant conditioning strategies increased appropriate eye contact and that the skill generalized to other environments. Mar and Sall (1995) used pretest and posttest measures and reported that engagement in social activities increased modestly for the three participants at school (range = 1–3 events) and somewhat greater at home (range = 2–6 events), though a social network analysis revealed that the students continued to have few friends who were consistently part of their lives. Neither study demonstrated experimental control, though both reports increase in the target social skills and opportunities.

3.3.4 *Adolescents*

One intervention study focused on the improvement of adolescents' social skills through the implementation of a specific curriculum. Kim (2003) employed a pretest/posttest control group experimental design and found no significant differences between the implementation of an assertiveness training curriculum designed to increase the social skills of teens and the control group as both groups showed improvement. Pre- and postmeasures included a standardized norm-referenced assessment of social skills that was completed by students, their parents, and their teachers; an assertiveness scale, a measure of cognitive distortion; and a role-play test.

4. DISCUSSION

4.1. Independent Living Skills

Three of the seven intervention studies reviewed demonstrated functional relations between intervention and all or some of the dependent variables of interest. All three addressed the construct of food management. While many of the included studies showed promising effects, in general, the single-subject experimental designs used by researchers did not have enough planned replications to show a functional relation.

4.1.1 *Food Management*

Physical prompts were more effective than verbal prompts in increasing a student's drinking skills (Farrenkopf et al., 1997). Researchers did not identify a criterion for mastery, a value used in alternating treatments designed to determine which intervention is more effective. Despite not having a set mastery criterion, the visual analysis showed 100% correct responses for three consecutive sessions in the physical prompting intervention, which is a commonly used mastery criterion. There were some points of overlap between the verbal and physical prompting interventions, but Farrenkopf et al. (1997) provided sufficient description of the circumstances related to that overlap and also continued the intervention phase until there was a clear delineation between the two interventions. After demonstrating the effectiveness of physical prompts at school, Farrenkopf et al. introduced the intervention at home and found similar results.

Sisson and Dixon (1986) did not plan enough replications in their design to make causal claims about whether praise and reinforcement can increase napkin and utensil use. Data were presented visually on a graph with three tiers, and while introduction of the intervention was time-lagged across tiers, there were different interventions for tier 1 (contingent reinforcement) and tiers 2 and 3 (praise and reinforcement). In addition, each tier had a different scale on the y-axis, making it impossible to compare data points across tiers. However, tier 1 had an ABAB design, allowing it to be analyzed separately. Contingent reinforcement was effective in increasing correct sitting posture. Although praise and reinforcement appear to have increased napkin use and utensil use, issues described earlier in the text and a lack of replications prevented any causal conclusions. Additionally, agreement data on procedural fidelity were collected on 35% of sessions, but there was low agreement for praise ($M=89.1\%$, range $=64$–97%). Considering praise was an integral component of the intervention, this low agreement on fidelity is an additional threat to the internal validity of the study.

One of the more surprising findings was that the removal of outside help could increase independence in a cafeteria line (Stolz & Wolf, 1969). While this idea seems very basic, it is worth reporting, especially because of reported expectations of students' continued dependence on others for independent living (Sacks et al., 1998). When help from aides was removed, the participant's independent retrieval of items in the cafeteria increased to 100% over 18 sessions. However, for their examination of eating behaviors, Stolz and Wolf (1969) used an AB design, which is not experimental. In addition, interobserver agreement was very low (47% for appropriate eating, 65% for appropriate but sloppy eating, and 80% for inappropriate eating). Although the visual data for inappropriate eating appear to show a decrease in behavior after the introduction of training, the design prevents any causal conclusions, and the low percentage of agreement does not instill confidence in the results.

Although Taylor's (1987) intervention appeared to be effective for all behaviors (i.e., preparing cereal, chocolate milk, and juice), there were no probe data immediately preceding the introduction of the intervention in tier 2 or tier 3. This omission prevents a true comparison from baseline to intervention, because the trend or level of behavior could have changed in the preceding sessions (25 sessions without a probe-preceded chocolate milk and 35 sessions without a probe-preceded juice).

Gumpel and Nativ-Ari-Am (2001) taught four students with visual impairment to shop for items from a grocery list using 17 chained skills identified by researchers. Two students received task analysis training, and two students received self-monitoring training. Task analysis training appeared to increase students' ability to purchase correct items but did not increase their ability to complete the skills in the correct sequence. Self-monitoring appeared to stabilize, but not increase, students' ability to purchase correct items or complete the correct sequence of skills. A multiple baseline design was used with each dyad, but multiple baseline designs require at least three tiers of replicated effects to show a functional relation between the intervention and a change in behavior. Even though task analysis appeared to increase correct independent purchases, there was not enough evidence to support a causal claim.

4.1.2 Personal Management (Oral Hygiene)

We can be confident that children with visual impairment have poorer overall oral hygiene than peers with sight (Ameer et al., 2012; Anaise, 1979; Ozdemir-Ozenen et al., 2012) and that there is a correlation between the amount of functional vision and amount of plaque (Anaise, 1979; Greeley et al., 1976). The design used by Shih and Chang (2005) did not provide enough opportunities for intrasubject replication. The first intervention included the entire package of instruction (classroom and coaching), and the second intervention phase was just coaching. All three participants had less plaque after intervention, but there was no change between the first and second phases of intervention or from intervention to the maintenance/return to baseline phase. It appears, then, that, after initial intervention with classroom instruction and coaching, no further coaching was necessary to maintain decreased plaque levels. However, more research is necessary to show that students can decrease their plaque levels through instruction in oral hygiene skills.

4.1.3 Money Management

A multiple baseline design requires at least three tiers (behaviors) to demonstrate a functional relation, and Hanney and Tiger (2012) had only two trained behaviors. Even though they collected data on four other behaviors, these tiers were not part of the time-lagged, multiple baseline design. Even though training appeared to increase students' correct responses and training appeared to generalize to untrained behaviors, there were not enough replications to make a causal claim.

4.1.4 Implications for Practice and Research

Overall, the research on independent living skills for students with visual impairment does not allow us to draw many causal conclusions. We can say that physical prompts are probably more effective than verbal prompts in increasing independent drinking skills of teenaged students with CVI at school or at home (Farrenkopf et al., 1997). There are, however, promising data that can be expanded or replicated using designs that plan for a sufficient number of replications. In fact, all seven interventions appeared to have positive effects on participants' behavior. The direction for future research on independent living skills, therefore, should not necessarily be focused on specific skills, because that will be driven by the needs of individual students. Instead, researchers need to use experimental single-subject designs that plan for enough replications (at least three) to demonstrate a functional relation.

4.2. Self-Determination

Four of the six intervention studies that were reviewed reported effective interventions; however, most did not provide sufficient information for readers to replicate their procedures.

4.2.1 Self-Awareness

Although clearly a medical, rather than an educational, intervention, Mohamed et al. (2011) demonstrated that 15 instructional sessions designed to increase knowledge of glaucoma and self-management of the condition were effective. The intervention was provided to dyads, with individual sessions for demonstration of skills such as administration of eye drops/ointment, use of compresses, and eye exercises. The questionnaires and behavioral observation scales described by the authors seemed to be carefully matched to the content of the intervention, and the authors managed to include 50 participants between 12 and 18 years, a relatively tight age range compared to some of the other studies.

Interestingly, inspection of descriptive statistics from Shapiro et al. (2005) suggested that females, not males, showed improvement in self-awareness of social acceptance, athletic competence, and physical appearance after participation in a 1-week athletic camp for students with visual impairments. The authors did not provide specific information about the strategies that were used to improve self-awareness of these concepts, however.

Alexander (1996) had only 20 participants in his intervention study designed to improve self-concept, and the small sample size may have accounted for the lack of impact. He used the Piers–Harris Children's Self-Concept Scale to measure self-concept pre- and postintervention. The intervention was five components from the Esteem Builders curriculum (Borba, 1991) considered appropriate for third to fifth graders: security, selfhood, affiliation, mission, and competence. Esteem Builders was designed as a 40-week program; Alexander's (1996) intervention was delivered in weekly sessions of unknown length for 18 weeks. It is possible that the Piers–Harris Children's Self-Concept Scale may not have assessed the constructs that were included in Esteem Builders. In addition, Esteem Builders delivered over 18 weeks instead of 40 might have resulted in a less than therapeutic dose of intervention. Finally, the author noted that 17 of the 20 students already had self-concept scores that were one standard deviation above the mean prior to intervention, leaving little opportunity to demonstrate changes in scores.

4.2.2 Goal Setting

Eniola and Adebiyi's (2007) intervention was reported to increase motivation to work in 32 students with visual impairments in Nigeria. They reported using lectures, demonstrations, discussions, homework, and outside activities to deliver the intervention with two sessions weekly for 6 weeks; however, they did not provide details about participants or content. The Work Value Inventory was used to assess the effectiveness of the intervention.

4.2.3 Self-Management

Jindal-Snape (2005b) provided specific information about target behaviors in her quest to teach self-evaluation/self-management skills, but she did not describe the process by which she taught the 9-year-old with total blindness to direct his eye gaze to sighted peers, nor did she describe the process by which she trained peers to provide feedback. Most importantly, she did not provide three replications of the impact of the intervention and so could not attribute the improvement in eye gaze and on-task behavior to the self-evaluation and peer feedback that were provided. Nevertheless, the results are promising and suggest that single-subject designs can be used effectively with the highly diverse group of students with visual impairments across cultures.

4.2.4 All Constructs of Self-Determination

As noted by Levin and Rotheram-Fuller (2011), measuring student perceptions of self-determination may not be the best way to assess whether students actually have skills related to self-determination. The global self-determination, self-concept, and self-perception of emotions rating scales, along with the relatively small sample size of 30, may account for the lack of statistical significance. It appears that data from parents, teachers, and students were entered as separate variables, further reducing statistical power. The descriptive statistics provided for mean pre- and postscores, however, seem consistent with the statistical tests, and so perhaps, the curriculum itself was not effective. As these authors acknowledged, self-ratings of perceptions of self-determination or instruction in self-determination may not be as valuable as opportunities to demonstrate self-determination in day-to-day, real-life situations.

4.2.5 Implications for Practice and Research

Given the constraints of the individual studies, we can draw relatively few conclusions about promoting self-determination of students with visual impairments. It appears that interventions that have a specific focus, such as improving awareness and management of glaucoma; increasing motivation to work; increasing self-awareness of social acceptance, athletic competence, and physical appearance; and increasing self-evaluation of behaviors that are usually reinforced visually, may be more effective than broad, global interventions to increase self-concept or self-determination. It appears that group interventions as short as one (Shapiro et al., 2005) to 6 weeks (Eniola & Adebiyi, 2007; Jindal-Snape, 2005b) can be effective in promoting self-determination, while some interventions of 15- to 18-week duration were ineffective (Alexander, 1996; Levin & Rotheram-Fuller, 2011).

Although a functional relation was not documented, data from Jindal-Snape's (2005b) single-subject intervention study suggested that individualized interventions that target specific needs of students could be effective. With careful attention to quality indicators of single-subject design, it is likely that we can design more interventions that will be effective in promoting self-determination.

4.3. Social Interaction Skills

Nine of the 10 intervention studies that were reviewed reported effective interventions: seven studies used a single-subject design, but only two of the seven had enough replications and experimental control to demonstrate a functional relation.

4.3.1 Preschoolers

In the oldest of the three studies on preschool-age children, Raver (1984) demonstrated that direct instruction and positive reinforcement of the target behavior provided by classroom teachers were successful at increasing head orientation in a 3-year-old. An unexpected finding was that, without instruction, peers began praising the child for looking at them, and the child even sought praise when she knew she was looking at the speaker. While this intervention showed experimental control and positive results, the generalizability of the findings is limited due to the lack of replication with other students.

Gronna et al. (1999) used puppets and sociodramatic scripts to train four same-age peers along with the subject on social behaviors including (a) greetings, (b) response to greetings, (c) response to conversation initiations, (d) verbal initiations of a participant-approached conversation, and (e) verbal initiations of a peer-approached conversation. There appeared to be a functional relation between the intervention procedure and an increase in the targeted social skills. However, generalizability of this finding is limited, as this study also involved only one participant. Based on these results, using puppets and modified published scripts with a structured learning approach can increase the prosocial behaviors of a preschool child who is visually impaired.

D'Allura (2002) compared an inclusive preschool classroom and a self-contained preschool classroom to explore the efficacy of a cooperative learning strategy for increasing peer interactions and group functioning. Students were not randomly assigned to classrooms, and the samples within each classroom were not equal, but an environmental analysis revealed that children with visual impairments in both classrooms spent comparable amounts of time in solitary play at the beginning of the study. D'Allura (2002) reported significant differences in the time spent interacting with peers and time spent in solitary play between the two classrooms: Children in the self-contained classroom actually increased their interactions with adults rather than peers, while children in the inclusive classroom increased their interactions to levels similar to sighted peers. While the implementation of the cooperative learning strategy appeared to have had an effect on the interactions with peers in the inclusive classroom, the design of the study did not control for all possible interacting variables, and students were not randomly assigned. It is unclear whether the implementation of the cooperative learning strategy or the presence of sighted peers or a combination of the two was responsible for the observed changes. Although the results of this study are promising and suggest that cooperative learning strategies with sighted peers may increase the number of interactions between students who are visually impaired and classmates, further research is needed to determine factors that contribute to children's social development and a more controlled research design should be used.

Although studies involving preschoolers suggest that the development of social skills in young children with visual impairment can be influenced through direct interventions, methodological problems with each of these studies reduce their generalizability. Replication with greater numbers of students and tighter experimental control are necessary before these interventions can be used with confidence. In particular, it would be interesting to see D'Allura's (2002) study conducted in equivalent integrated classrooms in which students with visual impairments are enrolled. Such a study could possibly provide insight into the value of carefully implemented cooperative learning strategies with preschoolers who are blind or who have low vision.

4.3.2 Children in Middle Childhood

All four studies of social skills of children in middle childhood used peer-mediated interventions. Sisson et al. (1985) trained two sighted peers to initiate play, offer assistance, and share materials with four students who were legally blind and had additional disabilities. Peers' initiations increased modestly after peer training, but social behaviors increased more substantially for all participants and generalization probes were mostly above baseline. A 4-month follow-up found maintenance of high levels of targeted social behaviors for students with and without disabilities. It should be noted that the peers in the study attended a different school than the target participants and were only in the same room together during play sessions. Perhaps more interactions and increases in social behavior would have occurred if all participants were enrolled in the same classroom. Though increases in social behaviors occurred for all participants, there were not enough replications of effect to demonstrate a functional relation, though the use of peers in social skills interventions appears to be a promising strategy.

Sacks and Gaylord-Ross (1989) compared peer-mediated and adult-directed interventions to determine which type was more effective at increasing social skills of children with visual impairment. Unlike other studies, Sacks and Gaylord-Ross (1989) recruited 15 students with visual impairments who were described as socially isolated but with no other disabilities. Students were assigned to one of three conditions: control, peer-mediated, or teacher-directed. Visual analyses of the data show larger gains in the teacher-directed group during training, though a statistical analysis found no significant differences, likely because of the small number of subjects involved. While targeted behaviors improved using both approaches, generalization to natural settings in the follow-up phase occurred only for students who were in the peer-mediated group. Because these researchers demonstrated sufficient experimental control, it appears that peer-mediated interventions lead to greater generalization of prosocial behavior than the teacher-directed intervention.

Two studies conducted by Jindal-Snape (2004, 2005a) sought to investigate the effectiveness of using self-evaluation to increase targeted social skills in a total of three children who were blind in India. In both studies, self-evaluation involved the participant setting criteria for the desired social behavior and then evaluating the use of the behavior after 2 min. Jindal-Snape (2004) compared the accuracy of self-evaluation with and without feedback of two female students. She found that adult feedback was critical to the accuracy of the self-evaluation and concluded that self-evaluation is most effective when its accuracy is confirmed, particularly for those skills that require visual cues. Demonstration of this finding with two students with varying social needs increases the confidence of this finding, which is also enhanced by the dramatic changes in level reported for both girls and the maintenance of these high levels at follow-up. However, a functional relation was not established due to less than three baseline data points and only two replications for each participant. In a similar study, Jindal-Snape (2005a) trained the participant to recruit feedback from two peers on the accuracy of his direction of gaze, but not to recruit feedback from peers for his conversation behaviors. Jindal-Snape (2005a) reported that the participant spontaneously asked peers for feedback on eye gaze during conversations, and his engagement in social interactions increased, as did his direction of gaze during these conversations and the accuracy of his self-evaluation of this behavior. Peers not only provided accurate feedback but also became more competent at providing information about the social environment. While there were general increases of target behaviors and better performance after feedback was received, this intervention was implemented with only one baseline session for direction of gaze and only three baseline sessions for conversation skills, which appeared to be trending upward. Stable baselines and longer interventions with more replication would strengthen the results and demonstrate a stronger relationship between the intervention and social skills outcomes.

4.3.3 Students with Deafblindness

Yarnall (1979) targeted eye contact in a child with visual impairment who was also deaf and who did not use any formal methods of communication. After 4 months of the intervention, the child made eye contact for more than 300 s during each of five consecutive 10-min classroom sessions. Generalization of the skill to other nonclassroom environments was reported, though not detailed in the published description of the study. Because the researcher used an AB design, experimental control was not demonstrated and a results must be interpreted cautiously.

Mar and Sall's (1995) intervention had three main components: frequent collaborative team meetings, educator support (consisting of the provision of informational resources, the promotion of adapted materials equipment, environments, in-service training, and modeling), and parent support (consisting of the provision of information, support for involving children in community activities, and networking with other parents). While the results indicated increases in social activities and social networks, it appeared that maintaining consistent friendships was difficult for all three subjects. The authors concluded that community-based program supports that are related to those being provided at school might be required to sustain the ongoing relationships of students who are deafblind. These findings have limited generalizability, due to the small number of participants and the lack of experimental control. They do, however, lend support to the notion that increasing the socialization of children with deafblindness is extremely difficult and that even complex interventions that involve interested parties in several environments may be insufficient.

4.3.4 Adolescents

Kim (2003) attempted to demonstrate experimental control by matching and randomly assigning 26 legally blind students to a treatment or control group. Kim (2003) used five different assessments to get parent, student, and observer ratings to measure social skills prior to and after the implementation of the intervention. Kim (2003) met with the treatment group at their school once a week for 12 weeks for a full period of 55 min. The results indicated that the intervention was not effective, though social skills improved in both groups, and there were almost statistically significant differences in the student-reported ratings between the two groups. Kim (2003) reported that the intervention might not have been intense enough as there was little incentive for the participants to work on skills or do homework away from the weekly class time, which she also felt needed to increase in frequency and duration.

4.3.5 Implications for Practice and Research

Despite the limitations of the individual studies, we can draw some noteworthy conclusions about facilitating the development of social interaction skills in children with visual impairments. Of perhaps utmost importance is the role peers can play and the influence they may have on the development of specific social behaviors. The multiple baseline studies conducted by Sisson et al. (1985), Sacks and Gaylord-Ross (1989), and Jindal-Snape (2005a) have suggested the important role of typical peers in the development of social skills within natural settings. These researchers proposed that this impact may be related to the similarity of stimuli when new skills are taught by the same peers with whom they are going to be used. Additionally, Raver (1984), Gronna et al. (1999), and D'Allura (2002) showed that peers can intentionally or unintentionally influence the development of social behaviors in preschool children and could possibly be as, or more, effective at increasing targeted social behaviors than adult-implemented interventions.

Jindal-Snape (2004, 2005a) had advanced the possible efficacy of self-evaluation/self-management of students with visual impairment, ages 8–11, to evaluate their own performance of specific social skills and, when appropriate, to recruit verification of that evaluation from peers. Not only targeted skills but also nontargeted social skills improve when this strategy was used. In addition, peers became more competent at providing information about the social environment. Further replications, however, are required to confirm this relationship.

The two studies (Mar & Sall, 1995; Yarnall, 1979) that focused on students with deafblindness highlight the heterogeneity of this population and the importance of selecting social skills interventions that address students' intellectual and communication levels and their visual and auditory capacities. Both studies reported that their interventions were effective and lend some support to the notion that increasing the socialization of children with deafblindness is extremely difficult; even complex interventions that involve interested parties in several environments may be insufficient. Mar and Sall (1995) noted the possible importance of community-based interactions, in addition to social interactions with friends at school and home, to establish and sustain friendships.

Studies that employed a single-subject design and targeted specific social skills appeared to be more effective at increasing social skills than group design studies that targeted more global social skills. Several studies showed the possibility of increasing eye gaze or eye contact in children with visual impairment or deafblindness through interventions that involve direct instruction and reinforcement (Raver, 1984; Yarnall, 1979), self-evaluation with feedback (Jindal-Snape, 2004, 2005a), or peers (Jindal-Snape, 2005a; Sacks & Gaylord-Ross, 1989). It was also possible to increase conversation and interaction skills through interventions, but some type of incentive or reinforcement may be necessary for students to recognize the importance and social consequences of better social skills.

5. RESEARCH PRIORITIES

This review of the literature, while extensive (though not necessarily exhaustive), has revealed that the intervention research conducted in the ECC areas of independent living skills, self-determination skills, and social interaction skills can only be described as paltry. In these three areas, each of which is thought to contribute substantially to the adult personal and career success of individuals with visual impairments, only 26 articles describing intervention research were identified. As reported here, many of these studies have methodological flaws that limit the ability of practitioners to use them as research-based interventions. Most, however, do reflect promising practices that should be researched further. Clarification of research questions, refinement of study designs, and replication of research that incorporates an increased number of subjects will be important as the field moves forward with its research agenda.

It is not surprising that most of the research that has been conducted in these areas of the ECC has involved descriptive nonexperimental methods, such as interviews or surveys, since conducting intervention research on students with visual impairments is fraught with challenges. The number of children whose learning is impacted by blindness or low vision in the United States is extremely small, with fewer than 100,000 students, aged birth to 22, suspected of having this disability nationwide (Kirchner & Diament, 1999a, 1999b). The low prevalence of this population, estimated between 0.1% and 0.2% of school-age students (Kirchner & Diament, 1999a, 1999b), makes conducting intervention research in the United States, where most students are provided educational services in local school districts, very difficult. Other barriers to research also exist, including the fact that this population of students is extremely heterogeneous. Heterogeneity exists in the level of visual acuity, visual functioning, age of onset, the presence of concomitant disabilities, previous educational experiences, school placement, and a host of other factors that influence students' behavior.

The low prevalence of students with visual impairments makes conducting research using single-subject research designs particularly attractive. In these interventions, subjects can be used as their own controls, and the designs allow for flexibility in that research can be conducted across environments, behaviors, subjects, and routines. Experimental control can be achieved through the use of designs that plan for at least three replications of an intervention effect (e.g., ABAB) and that incorporate generalization probes and maintenance phases across multiple baselines (three or more). Visual analysis of data permits the detection of changes in behavior that otherwise might not be identified through statistical analysis. External validity can be strengthened through multiple replications with similar subjects. A priority then for additional research in these areas is the use of single-subject designs that are replicated, across at least three participants or behaviors, to contribute to the development of a sound research base.

Regarding research design, outcome measures and measurement tools should be closely related to the intervention that is provided. Researchers who reported effective interventions tended to have outcome measures that were tailored to the intervention. Additionally, participants must be described in detail, and there should be sufficient numbers of participants to detect statistical significance in group study designs. Most authors acknowledged the heterogeneity of students with visual impairments, and yet they did not provide sufficient descriptions for us to know for whom the intervention was effective. Measures of central tendency were used to test most interventions, and the heterogeneity of students with visual impairments means that such measures may obscure findings for individual students. In some cases, researchers provided descriptive statistics that were valuable in understanding findings; all researchers need to provide descriptive statistics, and they need to examine results individually to see if outliers may be obscuring findings.

In addition, the research base could be increased through the collaboration of researchers in countries where the educational systems rely on grouping students with visual impairments in specialized schools for students with visual impairments. The large number of students enrolled in these schools could be included in studies using group designs. Random assignment to groups, matching subjects' characteristics, and alternate interventions could be employed. Statistical analysis of data could demonstrate efficacy of interventions.

The priorities, then, are for more and better intervention research. Hatlen (2007) quoted a colleague as describing the profession of teaching children with visual impairments as a "folk art." As part of this folk art, practices are handed down from one generation of teachers to the next. That there have been only 26 studies over the past 100 years that use research designs to explore the effectiveness of instructional practices in these three areas of the ECC provides evidence that practitioners and parents continue to use untested and unproven strategies to facilitate the development of these skills. Now is the time to stop exploring the unique characteristics of children who are blind and who have low vision and begin to discover practices that improve their functioning in the areas of social interaction, self-determination, and independent living.

REFERENCES

Agran, M., Hong, S., & Blankenship, K. (2007). Promoting the self-determination of students with visual impairments: Reducing the gap between knowledge and practice. *Journal of Visual Impairment & Blindness*, *101*(8), 453–464.

Alexander, F. E. (1996). Self-concepts of children with visual impairments. *RE:view*, *28*, 35–43.

Ameer, N., Durvasula, S., Neerudu, M., Palakuru, S., Palaparthi, R., & Singam, H. (2012). Oral hygiene and periodontal status of teenagers with special needs in the district of Nalgonda, India. *Journal of Indian Society of Periodontology*, *16*, 421–425.

Anaise, J. Z. (1979). Periodontal disease and oral hygiene in a group of blind and sighted Israeli teenagers (14-17 years of age). *Community Dentistry and Oral Epidemiology*, *7*, 353–356.

Blankenship, K. (2007). Effective instruction in the expanded core curriculum. In *Paper presented at the preconference day-long presentation, Texas Focus, San Antonio*.

Blankenship, K. (2008). Effective instruction in the expanded core curriculum. In *Paper presented at the New Mexico AER conference, Albuquerque.*
Borba, M. (1991). Esteem builders: A K-8 self esteem curriculum for improving student achievement, behavior and school climate. *Austin, TX: Pro-Ed, Inc.*
Brown, J. E., & Beamish, W. (2012). The changing role and practice of teachers of students with visual impairments: Practitioners' views from Australia. *Journal of Visual Impairment & Blindness, 106*, 81–92.
Cleveland, J., Clinkscales, R. M., Hefner, N., Houghtling, D., Kubacak, C., & Sewell, D. (2007). Becoming empowered through self-determination skills. *Texas School for the Blind and Visually Impaired. Austin, TX.*
Crudden, A. (2012). Transition to employment for students with visual impairments: Components for success. *Journal of Visual Impairment & Blindness, 106*, 389–399.
D'Allura, T. (2002). Enhancing the social interaction skills of preschoolers with visual impairments. *Journal of Visual Impairment & Blindness, 96*, 576–584.
Eniola, M. S., & Adebiyi, K. (2007). Emotional intelligence and goal setting — an investigation into interventions to increase motivation to work among visually impaired students in Nigeria. *British Journal of Visual Impairment, 25*(3), 249–253.
Farrenkopf, C., McGregor, D., Nes, S. L., & Koenig, A. J. (1997). Increasing a functional skill for an adolescent with cortical visual impairment. *Journal of Visual Impairment & Blindness, 91*, 484–493.
Field, S., Martin, J., Miller, R., Ward, M., & Wehmeyer, M. (1998). Self-determination for persons with disabilities: A position statement of the division on career development and transition. *Career Development for Exceptional Individuals, 21*, 113–128.
Greeley, C. B., Goldstein, P. A., & Forrester, D. J. (1976). Oral manifestations in a group of blind students. *ASDC Journal of Dentistry for Children, 4*, 39–41.
Griffin-Shirley, N., & Nes, S. L. (2005). Self-esteem and empathy in sighted and visually impaired preadolescents. *Journal of Visual Impairment & Blindness, 99*, 276–285.
Gronna, S. S., Serna, L. A., Kennedy, C. H., & Prater, M. A. (1999). Promoting generalized social interactions using puppets and script training in an integrated preschool: A single-case study using multiple baseline design. *Behavior Modification, 23*, 419–440.
Gumpel, T. P., & Nativ-Ari-Am, H. (2001). Evaluation of a technology for teaching complex social skills to young adults with visual impairments. *Journal of Visual Impairment & Blindness, 95*(2), 95–107.
Hanney, N. M., & Tiger, J. H. (2012). Teaching coin discrimination to children with visual impairments. *Journal of Applied Behavior Analysis, 45*, 167–172.
Hatlen, P. (1996). The core curriculum for blind and visually impaired students, including those with additional disabilities. *RE:view, 28*, 25–32.
Hatlen, P. (2007). Assisting the visually impaired: From folk art to profession. *AER Report, 24*, 9.
Hatlen, P., & Blankenship, K. (2010). Expanded core curriculum: Continued exploration of the problem. Unpublished manuscript.
Individuals with Disabilities Education Improvement Act, 20 U.S.C § 1400 *et seq*, 2004
Jindal-Snape, D. (2004). Generalization and maintenance of social skills of children with visual impairments: Self-evaluation and the role of feedback. *Journal of Visual Impairment & Blindness, 98*, 470–483.
Jindal-Snape, D. (2005a). Self-evaluation and recruitment of feedback for enhanced social interaction by a student with visual impairment. *Journal of Visual Impairment & Blindness, 99*, 496–498.
Jindal-Snape, D. (2005b). Use of feedback from sighted peers in promoting social interaction skills. *Journal of Visual Impairment & Blindness, 99*, 403–412.
Kim, Y. (2003). The effects of assertiveness training on enhancing the social skills of adolescents with visual impairments. *Journal of Visual Impairment & Blindness, 97*, 285–297.

Kirchner, C., & Diament, S. (1999a). Estimates of the number of visually impaired students, their teachers, and orientation and mobility specialists: Part 1. *Journal of Visual Impairment & Blindness, 93*, 600–606.
Kirchner, C., & Diament, S. (1999b). Estimates of the number of visually impaired students, their teachers, and orientation and mobility specialists: Part 2. *Journal of Visual Impairment & Blindness, 93*, 738–744.
Lane, K. L., Carter, E. W., & Sisco, L. G. (2012). Paraprofessional perspectives on promoting self-determination among elementary and secondary students with high-incidence disabilities. *Exceptional Children, 78*, 237–251.
Levin, D. S., & Rotheram-Fuller, E. (2011). Evaluating the empowered curriculum for adolescents with visual impairments. *Journal of Visual Impairment & Blindness, 105*, 350–360.
Lewis, S., & Iselin, S. A. (2002). A comparison of the independent living skills of primary students with visual impairments and their sighted peers. *Journal of Visual Impairment & Blindness, 96*(5), 335–344.
Lewis, S., & McKenzie, A. R. (2010). A comparison of competencies, roles, supervision, and training needs of paraeducators working with students with visual impairments in local and residential schools. *Journal of Visual Impairment & Blindness, 104*(8), 29.
Lohmeier, K., Blankenship, K., & Hatlen, P. (2009). Expanded core curriculum: 12 years later. *Journal of Visual Impairment & Blindness, 103*, 103–112.
Mar, H. H., & Sall, H. (1995). Enhancing social opportunities and relationships of children who are deaf-blind. *Journal of Visual Impairment & Blindness, 89*, 280–286.
McKenzie, A. R., & Lewis, S. (2008). The role and training of paraprofessionals who work with students who are visually impaired. *Journal of Visual Impairment & Blindness, 102*(8), 459–471.
Mohamed, E. A., Bayoumi, O. R., & Draz, S. F. (2011). Impact of an educational programme on knowledge, beliefs, practices and expectations about care among adolescent glaucoma patients in Cairo. *Eastern Mediterranean Health Journal, 17*, 960–968.
Obiakor, F. E., & Stile, S. W. (1989). Enhancing self-concept in students with visual handicaps. *Journal of Visual Impairment & Blindness, 83*, 255–257.
Ozdemir-Ozenen, D., Sungurtekin, E., Cildir, S., & Sandalli, N. (2012). A comparison of the oral health status of children who are blind and children who are sighted in Istanbul. *Journal of Visual Impairment & Blindness, 106*, 362–367.
Pugh, G. S., & Erin, J. (Eds.), (1999). *Blind and visually impaired students: Educational service guidelines.* Watertown, MA: Perkins School for the Blind.
Raver, S. (1984). Modification of head droop during conversation in a 3-year old visually impaired child: A case study. *Journal of Visual Impairment & Blindness, 78*, 307–310.
Sacks, S., & Gaylord-Ross, R. (1989). Peer-mediated and teacher-directed social skills training for visually impaired students. *Behavior Therapy, 20*(4), 619–640.
Sacks, S. Z., & Silberman, R. K. (2000). Social skills. In A. J. Koenig & M. C. Holbrook (Eds.), *Foundations of education: Volume II: Instructional strategies for teaching children and youths with visual impairments* (pp. 616–652). New York, NY: AFB Press.
Sacks, S. Z., Wolffe, K. E., & Tierney, D. (1998). Lifestyles of students with visual impairments: Preliminary studies of social networks. *Exceptional Children, 64*, 463–478.
Sapp, W., & Hatlen, P., (2007). Perceptions of teachers of students with visual impairments on their preparation to teacher the expanded core curriculum. Unpublished manuscript.
Sapp, W., & Hatlen, P. (2010). The expanded core curriculum: Where we have been, where we are going, and how we can get there. *Journal of Visual Impairment & Blindness, 104*, 338–348.
Self-Determination Technical Assistance Center. (2000). *Self-determination component skills*, Retrieved from http://www.sdtac.uncc.edu/.
Shapiro, D. R., Moffett, A., Lieberman, L., & Dummer, G. M. (2005). Perceived competence of children with visual impairments. *Journal of Visual Impairment & Blindness, 99*, 15–25.

Shih, Y., & Chang, C. S. (2005). Teaching oral hygiene skills to elementary students with visual impairments. *Journal of Visual Impairment & Blindness*, *99*, 75–80.

Sisson, L. A., & Dixon, M. J. (1986). Improving mealtime behaviors of a multihandicapped child using behavior therapy techniques. *Journal of Visual Impairment & Blindness*, *80*, 855–858.

Sisson, L. A., Van Hasselt, V. B. V., Hersen, M., & Strain, P. S. (1985). Peer interventions: Increasing social behaviors in multihandicapped children. *Behavior Modification*, *9*(3), 293–321.

Stolz, S. B., & Wolf, M. W. (1969). Visually discriminated behavior in a "blind" adolescent retardate. *Journal of Applied Behavior Analysis*, *1*, 65–77.

Taylor, R. G. (1987). Teaching a severely handicapped deaf-blind young woman to prepare breakfast foods. *Journal of Visual Impairment & Blindness*, *81*, 67–69.

Wolffe, K., Hatlen, P., & Blankenship, K. (2011). Report to the nation. Unpublished manuscript.

Wolffe, K., Sacks, S., Corn, A., Erin, J., Huebner, K., & Lewis, S. (2002). Teachers of students with visual impairments: What are they teaching? *Journal of Visual Impairment & Blindness*, *96*, 293–304.

Yarnall, G. D. (1979). Developing eye contact in a visually impaired, deaf child. *Education of the Visually Handicapped*, *11*, 56–59.

CHAPTER EIGHT

Orientation and Mobility for Students with Visual Impairments: Priorities for Research

Robert Wall Emerson[*,1], **Tessa McCarthy**[†]
[*]Department of Blindness and Low Vision Studies, Western Michigan University, Kalamazoo, Michigan, USA
[†]Department of Education, North Carolina Central University, Durham, North Carolina, USA
[1]Corresponding author: e-mail address: robert.wall@wmich.edu

Contents

Abstract

This chapter broadly reviews the development of the field of orientation and mobility (O&M) and its application to children and then looks more closely at several current areas of research in O&M. Research studies on the topics of quiet cars, complex intersections and the built environment, electronic travel aids, long canes, children with multiple disabilities, and instructional techniques in general are reviewed and synthesized. Future directions of research in the field such as cross-disciplinary cooperation, providing people in developing countries access to independent travel, and technology trends are discussed. The summary point is that the field of O&M needs to use research to build and validate its knowledge base, expand its partnerships with other fields, and build capacity in the field.

International Review of Research in Developmental Disabilities, Volume 46
ISSN 2211-6095
http://dx.doi.org/10.1016/B978-0-12-420039-5.00008-3

1. INTRODUCTION

The field of orientation and mobility (O&M) has traditionally been geared toward services provided for adults (Wiener & Sifferman, 2010). The history of O&M services for children is short and was dominated in its beginning by adapting and applying techniques and tools developed for adults to use for children. As more O&M professionals worked with children, the fundamental differences in teaching adults and children became more evident: especially in that children require more guided concept development, more "hands on" experience with their environment, more time to generalize concepts and skills, and often acquire skills in a different sequence than adults (Fazzi & Naimy, 2010; Skellenger & Sapp, 2010).

While the teaching of adults and children is very different, the core set of O&M principles is the same: giving an individual the concepts, skills, and tools necessary to move efficiently through their environment. While this may look very different when applied to adults and children, the environments in which movement is occurring are often very similar. Both adults and children must navigate homes and neighborhoods, schools and office buildings, shops, and streets. While an adult is expected to demonstrate higher levels of independence than a child, both must acquire understanding of these environments and how to navigate them.

Because the modern field of O&M was developed to service adults, research conducted to develop the field of O&M has tended to be dominated by the development of skills and tools for teaching adults. And, because adults typically do not need the level of concept development and complex layering and scaffolding of skills that children need, much of this research has focused on basic skill teaching instead of how to meld O&M instruction with other skill and concept areas. While some research into O&M techniques, tools, and instructional practices can be applied to both adults and children, since the 1970s, there has been a slowly growing stream of research designed to address topics relevant only to children. The provision of O&M services for children has developed more slowly than those for adults but has gained momentum in the last few decades. This may partially be due to the fact that the majority of O&M instructors primarily teach children. In 2000, Wiener and Sifferman found that 46.9% of surveyed certified O&M instructors taught in public or residential schools. When questioned specifically about the age group of clients served, 36.5% of all respondents spent some of their time teaching toddlers, 67.7%

of all respondents spent some of their time teaching children aged 4–12 years, and 67.2% of all respondents spent some of their time teaching adolescents aged 13–17 years (Wiener & Sifferman, 2000).

This chapter will outline the development of the modern field of O&M and discuss significant research efforts that have influenced the field. More recent research efforts that have the potential to significantly impact the field will be discussed, followed by thoughts on future directions the field of O&M might take.

2. BACKGROUND AND RATIONALE FOR REVIEW

People who are blind have been traveling with the aid of walking sticks for millennia but with no cohesive system of techniques or standard mobility tool. If a person who was blind received any training in techniques for getting around, it was generally because another person who was blind was in the area and communicated some techniques gained through experience. The first school for children who were blind was opened in France in 1784 (Hatlen, 2000). In the United States, Perkins School for the Blind (originally called the New England Asylum for the Blind) opened in 1829, and many states had some sort of school for children who were blind by the late 1870s (Hatlen, 2000). However, the curriculum at these schools often did not include instruction in independent mobility. This is largely due to the difference between teachers of students with visual impairments (TVIs) and O&M specialists. TVIs focus on making adaptations and accommodations to provide students who are blind and visually impaired access to the same educational materials and experiences as typically sighted peers in addition to teaching certain skills that are specific to children who are blind and visually impaired. We refer to these disability-specific skills as the expanded core curriculum (ECC). Orientation and mobility specialists focus on one area of the ECC, O&M. Orientation and mobility specialists teach students to move, understanding spatial relationships and updating these relationships as an individual travels, as well as how to travel through space safely and efficiently with the use of devices such as a long cane (Wiener, Welsh, & Blasch, 2010).

The modern field of O&M began in the 1940s in the Veterans Administration (VA) hospital system. Some people in the VA system determined that veterans who had lost some or all of their vision in the Second World War needed training in how to get around. Dr. Richard Hoover developed the modern white cane, based on a style developed in England in the 1920s and championed in the United States in the 1930s by Lions Club International (New Jersey Council of the Blind, n.d.) and laid out a set of basic techniques for using the long cane. Armed with the new canes and techniques, a set of occupational therapists at the Hines VA became the first official O&M instructors and began teaching veterans on techniques for independent mobility (http://www.acb.org/tennessee/white_cane_history.html).

The Hines VA hospital in Chicago was the center of O&M development and instruction for many years, until the first university-based training programs began in the early 1960s. While the modern field of O&M was developing in these early years, the focus was on adults. Because of this focus, many in the field believed that children were not capable of using the long cane appropriately. Indeed, many advocated that children not be taught how to travel with a long cane until they were around 10 years old (Pogrund & Rosen, 1989).

Research conducted by Fraiberg in the 1960s began a line of thought that indicated that the development of children who were blind might be impeded if they did not go through the proper sequence of motor skill acquisition when they were young (Fraiberg, 1971). While Fraiberg did not expressly advocate for O&M training, the scene was set for others to develop the idea. In 1967, two seminal articles were published regarding O&M with young children: *Orientation and mobility for preschool blind children* (Eichorn & Vigaroso, 1967) and *Developmental concepts of blind children between the ages of three and six as they relate to orientation and mobility* (Hapeman, 1967). In the 1970s and 1980s, other researchers developed more techniques for working with children on developing independent and effective travel. The understanding that O&M, when discussed in the context of children, was more than using a long cane to get from point A to point B began to develop from this research. Orientation and mobility for children began to involve development of body image, understanding of spatial concepts, and understanding how their own movement affected the relative position and direction of other objects. These developments were summarized in 1979 (Ferrell, 1979), and the Peabody Preschool O&M Project was funded in 1984 to investigate the topic of O&M with young children (Hill, Rosen, Correa, & Langley, 1984). The final report for this demonstration project was released in 1988 (Hill, 1988), just as a surge of curricula and articles about cane design and adaptive teaching techniques regarding O&M with young children were being published (e.g., Bosbach, 1988; Brown & Bour, 1986; Clarke, 1988; Foy, Kirchner, & Waple, 1991;

Hill, Dodson-Burke, & Smith, 1989; Kronick, 1987; O'Mara, 1989; Pogrund & Rosen, 1989).

Special forms of the long cane were developed for use with young children or children with multiple disabilities (Bosbach, 1988; Foy et al., 1991), sometimes with the belief that young children were not physically able to wield the long cane appropriately (Mancil, Manuel, Sifferman, & Blasch, 1998). Originally called "precane" devices, these alternative mobility devices were very popular in the 1980s and 1990s (e.g., Clarke, Sainato, & Ward, 1994) and were gradually used with elderly people with vision loss as well as with children (Mancil et al., 1998). It is not uncommon now to have children who are blind using a cane (designed for their height) as soon as they are able to walk.

This brief review of O&M instruction and its relationship to children illustrates how research has been intertwined with development in the field. Given this, we wanted to look at recent research in the field of O&M to see how the field is currently changing by examining some of the more frequent topics published and presented at national conferences in recent years and then determine the directions we see the field moving in the future. While these topics may not be specific to children, shifts in the field of O&M impact how services are provided to children. The fact that O&M services for children now make up a significant part of the field means that trends in the field in general are also trends in services for children.

3. METHOD

A literature search was made using the terms "O&M" or "wayfinding" combined with "blind" or "deafblind" or "deafblind" or "deafblind" or "eye disorders" or "partially sighted" or "vision disorders" or "visual disabilities" or "blindness" or "visual impairment" or "visual impairments" or "visually impaired" or "partial vision." The date range was restricted to 2000 to the present and databases used included ERIC, PsycINFO, psycARTICLES, and ProQuest education journals. We also reviewed topics of O&M-themed presentations made at the International Mobility Conference in 2003, 2006, 2009, and 2012; Association for Education and Rehabilitation of the Blind and Visually Impaired (AER) International in 2008 and 2010; the International Low Vision Conference in 2005 and 2008; and planned presentations at the AER O&M Conference in 2013. Repeats of articles or presentations from the resulting lists were eliminated.

Each presentation and article was characterized by a single primary category and, when appropriate, by a secondary category. The list of categories was created as the list was reviewed so that each new article or presentation that did not fit a previous category was given a new category. In the end, 181 articles from 37 journals (the three most represented were *Journal of Visual Impairment & Blindness*, *International Journal of Orientation and Mobility*, and *RE:view*) and 541 presentations from the nine conferences were categorized. The list of categories used and their relative frequencies is shown in Table 8.1. Categories that appeared to fall within a broader domain (such as "built environment" and "roundabouts") were combined to obtain a more robustly representative category.

From this table, and especially after grouping similar categories, the most common categories that also had the most potential relevance to children were selected for a more thorough review. This included 9 of the 11 most frequent single categories. The category of "O&M profession" was broad and vague. Articles and presentations that went into "O&M profession" were discussions of how to train O&M instructors, needs of the field, issues in the field, Medicare billing, and licensure. It was decided that this was not one of the more relevant categories for children. After combining related topics, the final list of O&M categories with the most relevance to children included the built environment (including roundabouts, quiet cars, and accessibility), electronic travel aids (including GPS), cane use and design, O&M teaching focus (including street crossings), multiple disabilities (including deafblind), and early childhood.

In these six topical areas, a more targeted search of research and related articles was performed. A search of 49 article databases including dissertations, periodicals, newspapers, psychology journals, medical journals, engineering journals, and social science journals was conducted for each of the six topical areas. In order to restrict the results to only the most recent publications in each area, results were accepted from 2009 to the present. While an effort was made to gather an exhaustive list, in each topical area, there may be some articles that were missed or omitted due to a lack of direct relevance.

Some articles of particular importance published before 2009 were included in some of the topical areas. These articles were included because they either were data-based articles in an area that had few such or represented seminal articles that first introduced the topic in a general manner. In Section 4.1, three articles from 2005 and one from 2006 were included; in Section 4.2 one article from 2004 and three from 2007 were included that were seminal data-based articles in GPS use, and in Section 4.3, two articles

Table 8.1 Topic Categories and Frequencies Identified During the Literature Review Process

Category	Total Frequency	Presentation Frequency	Article Frequency	Combined Categories
O&M profession	68	51	17	
Teaching focus	60	50	10	+Street crossing (30) = 90
Dog guides	57	42	15	
Built environment	48	44	4	+Roundabouts (9) + accessibility (16) + quiet cars (7) = 80
International issues	33	29	4	
Multiple disabilities	32	27	5	+Deafblind (14) = 46
Electronic travel aids (ETAs)	31	20	11	+GPS (27) = 58
Street crossing	30	29	1	
Global positioning systems (GPSs)	27	22	5	
Cane use	25	17	8	+Cane design (9) = 34
Early childhood	22	19	3	
Low vision	20	18	2	
Assessment	20	12	8	
Technology	18	17	1	
Auditory skills	17	13	4	
Virtual reality	16	4	12	
Public transport	16	13	3	
Accessibility	16	15	1	
Tactile maps	15	10	5	
Neurology	14	5	9	

Continued

Table 8.1 Topic Categories and Frequencies Identified During the Literature Review Process—cont'd

Category	Total Frequency	Presentation Frequency	Article Frequency	Combined Categories
Deafblind	14	10	4	
Recreation	11	7	4	
Driving	10	9	1	
Cane design	9	5	4	
Roundabouts	9	8	1	
Perception	9	3	6	
Social aspects	9	4	5	
Echolocation	8	5	3	
Spatial perception	7	1	6	
Quiet vehicles	7	5	2	
Posture/gait	6	6	0	
Interagency issues	5	2	3	
Wayfinding	6	0	6	
Safety	5	4	1	
Indoor	5	2	3	
Elderly	5	4	1	
Family	4	2	2	
Math	2	1	1	
Concepts	2	0	2	
Orientation	2	2	0	
Employment	2	1	1	
Rural issues	1	1	0	
Client characteristics	1	0	1	
Music	1	0	1	
Physical education	1	1	0	

from 2005 were included that provided data-based evaluation of long cane characteristics.

The resultant articles were then reviewed for the following discussion of each of the topics to determine trends in these "hot topics." While this chapter focused on peer-reviewed research in O&M, the authors felt case studies, literature reviews, and practice pieces needed to be included to demonstrate the full range of topics that are currently important to the field of O&M. In several topic areas, such as the built environment, electronic travel aids, and long cane use, almost all relevant articles dealt with adults. However, these topics were preserved as they reflect current trends. As we noted in Section 1 to this chapter, new knowledge in the field of O&M often proceeds from investigations with adults to applications for children.

4. RESULTS

4.1. Complex Intersections and the Built Environment (Including Quiet Cars)

One of the fundamental aspects concerning O&M, whether for adults or children, is the consideration of the environment in which a person with a visual impairment will be traveling. While many indoor environments share common features such as door sizes, staircases, and hallways, outdoor environments include a very large range of features and variations within different features. In this topic area, 27 articles were identified and located. Five articles from before 2009 were included in this tally (four from 2005 and one from 2006) as they represented initial comprehensive treatment of this topic within the professional blindness literature. Of these five articles, four reported on practices or reviewed literature (Barlow & Franck, 2005; Long, Guth, Ashmead, Wall Emerson, & Ponchillia, 2005; Sauerbuger, 2006; Sauerburger, 2005), and one reported on research (Barlow, Bentzen, & Bond, 2005). Of the remaining 22 articles, 20 reported on research (Ashmead et al., 2012; Bourquin, Wall Emerson, & Sauerburger, 2011; Garay-Vega, Hastings, Pollard, Zuschlag, & Stearns, 2010; Geruschat, Fujiwara, & Wall Emerson, 2011; Gross, Lyon, Persaud, & Srinivasan, 2013; Guth, Long, Wall Emerson, Ashmead, & Ponchillia, 2013; Hassan & Massof, 2012; Hastings, Pollard, Garay-Vega, Stearns, & Guthy, 2011; Hong, Cho, & Ko, 2013; Kim, Wall Emerson, Naghshineh, Pliskow, & Myers, 2012a, 2012b; Ping, Diew, Menow, & Palanusamy, 2012; Robart & Rosenblum, 2009; Sanchez & Saenz, 2010; Schroeder et al., 2011; Scott et al., 2011a, 2011b; Wall Emerson, Kim,

Naghshineh, Pliskow, & Myers, 2011, 2013, Wall Emerson, Naghshineh, Hapeman, & Weiner, 2011), one described an analysis on a national dataset (Hanna, 2009), and one reported on practices (Strickfaden & Devlieger, 2011). All of the articles reporting results of research used adults as participants, although several either used intersections as the unit of analysis (Gross et al., 2013) or made use of larger datasets as well as collecting new data (Garay-Vega et al., 2010; Hastings et al., 2011).

Urban travel environments have changed substantially in the last few decades. Streets and intersections have grown larger, traffic signals are more coordinated, corners are more rounded, and traffic signals are more commonly actuated (Barlow et al., 2005). These changes in the urban travel environment have dramatically changed how a pedestrian with a visual impairment needs to approach the crossing of streets (Sauerbuger, 2006; Sauerburger, 2005). This not only often involves the use of accessible pedestrian signals and other technologies (Barlow & Franck, 2005; Ping et al., 2012) but also often requires paying attention to different cues from traffic (Long et al., 2005; Sauerbuger, 2006). At roundabout intersections, the lack of parallel and perpendicular traffic forces pedestrians to make crossing decisions either in a gap in traffic or when traffic is yielding (Guth et al., 2013). However, potential masking sounds from traffic in the circulating roadway can make the performance of these tasks difficult, especially in times of high traffic volume (Guth et al., 2013). Although crossing at roundabouts and channelized turn lanes can require different strategies than at signalized intersections, there are engineering and instructional approaches that can reduce the risks (Schroeder et al., 2011). In general, the needs of pedestrians, including pedestrians with disabilities, are becoming more of a consideration when designing travel environments such as converting intersections to roundabouts (Gross et al., 2013).

The impact of masking sounds, or any significant background noise levels, is highlighted when pedestrians must make crossing decisions in the presence of quiet vehicles. The increase in the number of hybrid and electric vehicles has the potential to significantly change the access pedestrians with visual impairments have to useful sound information (Garay-Vega et al., 2010; Hanna, 2009; Robart & Rosenblum, 2009; Wall Emerson, Kim, et al., 2011; Wall Emerson et al., 2013). Reducing the sound level of vehicles can make it more difficult to identify where a vehicle is going (Ashmead et al., 2012; Kim et al., 2012b), whether a vehicle is idling nearby (Hong et al., 2013; Wall Emerson, Kim, et al., 2011), whether a vehicle is starting up from a stop (Kim et al., 2012b; Wall Emerson,

Naghshineh, et al., 2011), or how well crossable gaps in traffic can be detected (Wall Emerson et al., 2013). The most common solution being explored to address these safety issues is to add artificial sounds to hybrid and electric vehicles. Guidelines for artificial sounds being developed by the National Highway Transportation Safety Administration (Hastings et al., 2011; Pedestrian Safety Enhancement Act, 2010) are informed by studies that have found beneficial results from adding such sounds (Kim et al., 2012a; Wall Emerson et al., 2013).

Given that urban travel environments have become more complex and the potential for useful acoustic information to be decreased, there is a need for strategies that maximize pedestrians' ability to optimize crossing safety. In situations where a yield or gap in traffic must be identified, there are strategies that offer a higher potential for yielding (Bourquin et al., 2011) or for persons with low vision to identify gaps (Geruschat et al., 2011; Hassan & Massof, 2012).

One feature of the urban travel environment that does not require access to acoustic information is tactile information on the walking surface. While truncated domes are becoming a common occurrence on wheelchair ramps at crossing points, they are designed only to provide awareness of the boundary between the walking surface and the road surface. Recently, however, there has been increased interest in how detectable warnings can be modified to provide alignment information for maintaining straighter street crossings (Scott et al., 2011a, 2011b).

These detectable warnings have also been used on platforms in metro systems. The urban metro system can be a very complicated travel system. Because these systems were designed to move many people on a daily basis, and to be useful as a means of daily travel, many people with disabilities may use these systems. However, when the earliest of these systems were built in the late 1800s and early 1900s, accessibility for pedestrians with disabilities was not a priority (Strickfaden & Devlieger, 2011). Urban metro systems are now often collaborating with people with different disabilities to optimize accessibility of redesigned or newly designed systems (Sanchez & Saenz, 2010; Strickfaden & Devlieger, 2011).

4.2. Electronic Travel Aids (Including Global Positioning System)

In this topic area, 24 articles were identified and located. Of these articles, 15 reported on research of some type (Havik, Kooijman, & Steyvers, 2011; Havik, Steyvers, van der Velde, Pinkster, & Kooijman, 2010; La Grow et al.,

2009; Lahav & Mioduser, 2004, 2007; Lahav, Schloerb, & Srinivasan, 2012; Lancioni, Singh, et al., 2010; Ponchillia, MacKenzie, et al., 2007; Ponchillia, Rak, et al., 2007; Roentgen, Gelderblom, & de Witte, 2011; Ross & Kelly, 2009; Sáenz & Sánchez, 2010; Simonnet & Vielledent, 2012; Wong & Tan, 2012; Wood, 2011), three reported on practices (Delgado Greenberg & Kuns, 2012; Penrod, Simmons, Bauder, & Brostek Lee, 2009; Phillips, 2011), four combined reporting on practices or doing a case study with the development of a new device (Hoefer, 2011; Kumar, Patra, Mahadevappa, Mukhopadhyay, & Majumdar, 2013; Roentgen, Gelderblom, & de Witte, 2012; Sánchez & Tadres, 2011), and two were literature reviews (Roentgen, Gelderblom, Soede, & de Witte, 2008; Smith & Penrod, 2010). Of the articles reporting results of research, 11 used some form of group experimental design, while four reported case studies (Lahav et al., 2012; Sáenz & Sánchez, 2010; Sánchez & Tadres, 2011; Wong & Tan, 2012), and two used a single-subject design (Lancioni, Singh, et al., 2010; Ponchillia, Rak, et al., 2007). Almost all of the research was conducted using adults as participants. Only two articles reported data on children or students (Phillips, 2011; Sáenz & Sánchez, 2010).

Orientation and mobility, whether with adults or children, has seen a dramatic increase in the use of technology in the last few years. Increased precision and lower cost of GPS devices have sparked a tremendous upsurge in the use of these devices by people with visual impairments. One survey identified 13 devices aimed at obstacle detection and orientation and 10 devices aimed at navigation that were available for travelers with visual impairments (Roentgen et al., 2008). Technology that uses GPS either in a standalone unit or through an application on a smart phone has greatly increased the amount and kind of information that persons with visual impairments can access. In general, the articles in this topical area could be classed as developing new technologies such as virtual reality training, evaluating GPS devices, or discussing how to teach the use of devices.

There has been a long history of electronic travel aids development for people with visual impairments (Smith & Penrod, 2010), and the recent past is no different. Lahav et al. (2012) published a case study using a system called the BlindAid, a continuation of a line of research investigating the use of virtual reality as a training tool for O&M performance and comparing performance in similar virtual and physical spaces (e.g., Lahav & Mioduser, 2004, 2007). A similar paradigm was used by Simonnet and Vielledent (2012) who assessed the ability of six blind sailors to use virtual maps to navigate a virtual haptic and auditory space. Both research groups discuss exploration strategies used by participants as well as how technology might be used to teach O&M skills.

A number of researchers investigated how well participants could perform tasks when using certain GPS devices. Ponchillia, MacKenzie, et al. (2007) and Ponchillia, Rak, Freeland, and LaGrow (2007) used a BrailleNote GPS to assess participants' ability to find a target and reorient within a typical travel environment and found that participants were quite accurate. Havik et al. (2010) assessed the Trekker and showed improved O&M skills and attitude to travel, while Wood (2011) investigated use of the Trekker Breeze. Roentgen et al. (2011) compared four devices (BrailleNote, Trekker, Trekker Breeze, and Wayfinder Access) in both operational ease and performance of O&M tasks. They reported that while newer technologies often enhance the performance of mobility tasks, their use is generally not as smooth and seamless as many users would like. User perceptions of GPS devices were explored by La Grow et al. (2009), while Havik et al. (2011) found that user preferences were often not consistent with the GPS characteristics that gave best performance.

Ross and Kelly (2009) addressed the issue of developing assistive navigation technology for indoor situations where traditional GPS technologies do not work by using "Cyber Crumbs" to provide additional information to assist participants who were blind when traveling in indoor environments. Roentgen et al. (2012) focused more on developing indoor environment where electronic devices might be evaluated. Sánchez and Tadres (2011) assessed Augmented Reality Tags for Assisting the Blind that use audio representation to support indoor navigation. Sáenz and Sánchez (2010) used a pocket personal computer device to collect data about a user's position and relate it to a representation of the indoor space on another device. A more specific indoor task was investigated by Lancioni, Singh, et al. (2010) where four participants, two ambulatory and two in wheelchairs, had to find specific doorways through verbal and vibratory feedback given via portable technology.

In addition to these orientation technologies, new technology is being used to develop enhanced methods to avoid obstacles and hazards. Kumar et al. (2013) used a system of ultrasonic sensors and USB cameras to detect obstacles via sonar and give audio information about objects to users, enhanced by information from a webcam. The viability of an open-source navigation device, based on sonar technology, was discussed by Hoefer (2011).

Smartphone applications have seen a dramatic increase in use recently, with many applications being designed specifically for users with visual impairments. In addition to using the camera on a smartphone to identify objects, read barcodes, identify colors, and even recognize faces, GPS-based applications can help individuals with visual impairments to locate nearby stores and restaurants, identify the distance to the nearest intersection, and locate nearby streets. Wong and Tan (2012) found numerous potential pitfalls in teaching the use of such smartphone applications to a user who was blind.

To be used efficiently and effectively, new technologies often require a substantial investment of time on the part of both professionals and consumers to understand and use well. This realization, no doubt, led Delgado Greenberg et al. (2012) to develop an approach and curriculum for teaching the use of GPS devices. Other tips and strategies have been offered by Phillips (2011), while Penrod et al. (2009) discussed the use of electronic travel aids in conjunction with a primary mobility system. Penrod et al. (2010) also discussed the effectiveness of a course designed to train instructors on how to teach electronic mobility aids.

4.3. Cane

In this topic area, 16 articles were identified and located. Of these articles, 9 reported on research of some type (Huang, Leung, & Wang, 2010; Kim & Wall Emerson, 2012; Kim, Wall Emerson, & Curtis, 2009, 2010a, 2010b, 2010c; Rodgers & Wall Emerson, 2005a, 2005b; Scott, 2010), three reported on practices (Penrod, 2012; Sauerburger & Bourquin, 2010; Zijlstra, Ballemans, & Kempen, 2013), two combined reporting on practices with the development of a new device (Bousbia-Salah, Bettayeb, & Larbi, 2011; Willins, 2012), and two were literature reviews (Ballemans, Kempen, & Zijlstra, 2011; Smith & Penrod, 2010). Of the articles reporting results of research, 8 used some form of group experimental design, while one reported a case study (Scott, 2010). Apart from Scott (2010), all of the articles had an adult focus, either in the use of participants or in the discussion of practices.

There has been a long tradition of developing new mobility devices for travelers who are blind, by either using new technology for orientation and navigation, modifying the long cane for enhanced obstacle detection, or adding a secondary device to cane use to enhance obstacle detection (Smith & Penrod, 2010). However, the long white cane has remained the most consistent mobility aid for travelers who are blind. While certain modifications such as jointed segments, newer materials, and a variety of tips have altered aspects of the long cane, the basic design has remained relatively stable since it was developed by Richard Hoover in the 1940s (Smith & Penrod, 2010). Even though the long cane was developed for use by adults about 70 years ago, and then adapted for use by children, there has been a lack of consistent research into the optimal properties of the long cane or a quantification of how certain techniques impact performance.

In the timespan defined for this chapter, there continued to be developments in cane design (e.g., Willins, 2012) and the addition of enhanced sensors to long canes (Bousbia-Salah et al., 2011). More frequently, however, authors examined how current cane characteristics impacted performance of O&M tasks. Researchers described the effects of weight (Rodgers & Wall Emerson, 2005a), length (Huang et al., 2010; Kim & Wall Emerson, 2012; Rodgers & Wall Emerson, 2005a), material (Rodgers & Wall Emerson, 2005b), and cane tip (Kim et al., 2010a) on the performance of O&M tasks such as drop-off detection, balance, and fatigue. Other authors looked at the impact of cane technique (Kim et al., 2009, 2010b) and user characteristics (Kim et al., 2010c) on O&M skill performance. In addition to these studies on cane and technique characteristics, other authors were interested in defining optimal procedures for teaching cane use in general (Penrod, 2012; Sauerburger & Bourquin, 2010) and specifically for children (Scott, 2010) or people with low vision (Ballemans et al., 2011; Zijlstra et al., 2013).

4.4. Teaching Focus, Including Street Crossings

In this topic area, 19 articles were identified and located. Three articles and two unpublished dissertations were identified but could not be located. Of the 19 articles that were located, two reported on research (Ross & Kelly, 2009; Wright & Wolery, 2014), one described a secondary analysis on national survey data (Anderson, 2010), four shared the results of case studies (Lahav et al., 2012; O'Mea, 2013; Sáenz & Sánchez, 2010; Sánchez & Tadres, 2011), one synthesized the literature (Wright, Harris, & Sticken, 2010), and 11 reported on practices (Barrella et al., 2011; Breslauer, 2010; Erin, 2009; Franck, Haneline, & Farrugia, 2011; Martinez, 2011; Mbulaheni, Dzapasi, & Sabeya, 2012; Penrod, 2012; Sapp, 2011; Sauerburger & Bourquin, 2010).

As previously mentioned, the research literature in O&M has historically focused on skills and tools for teaching adults. However, the recognition that children must be explicitly taught concepts and be provided with scaffolding to be effectively taught is reflected in the current literature. The research reports on this topic area were divided in terms of studying children versus adults. Of the two papers that reported on research, one used a quasiexperimental group approach with adults (Ross & Kelly, 2009), and one used a single-subject approach with children (Wright & Wolery, 2014). Likewise, one case study focused on working with adults (Lahav et al., 2012), and one case study focused working with children (O'Mea, 2013). The secondary analysis of survey data focused on elementary-aged children (Anderson, 2010), and the literature synthesis focused on students under the age of 21 (Wright et al., 2010). Two of the case studies included participants who were both adults and children (Sáenz & Sánchez, 2010; Sánchez & Tadres, 2011). Two of the practice reports likewise included implications for both adults and children (Erin, 2009; Mbulaheni et al., 2012), while one focused on interventions for children (Sapp, 2011). So, it appears that the literature on teaching O&M over the past five years reflects an equal interest in children and adults.

Many of the articles did not specify the level of vision of participants or if the practices were intended for individuals with low vision or total blindness. Of those that did specify, two studies included participants with light perception or less (Sánchez & Tadres, 2011; Wright & Wolery, 2014). The other two articles that described the vision of participants included individuals who were totally blind and individuals with low vision (Ross & Kelly, 2009; Sáenz & Sánchez, 2010).

Several themes related to teaching O&M surfaced. Some articles focused on strategies and methods for teaching O&M such as behavioral techniques (O'Mea, 2013; Wright & Wolery, 2014), new approaches to using resources in the natural environment (Feinberg, 2011), ways to make instruction fun (Sapp, 2011), and step-by-step guides (Penrod, 2012; Sauerbuger, 2006; Sauerburger, 2005; Sauerburger & Bourquin, 2010). One paper focused on who can teach O&M, specifically O&M instructors who themselves use a dog guide (Breslauer, 2010). Three papers focused on low-tech tools for teaching O&M such as maps and models (Erin, 2009; Wright et al., 2010) and canes (Sauerburger & Bourquin, 2010). Meanwhile, the majority of the studies and papers looked at providing high-tech tools such as transponders that provide auditory or tactile feedback to student travelers (Ross & Kelly, 2009; Sáenz & Sánchez, 2010; Sánchez & Tadres, 2011) and virtual reality or virtual environment as teaching tools (Lahav et al., 2012). Some publications focused on the lack of resources available for providing O&M instruction (Barrella et al., 2011); a paucity of resources for specific groups, such as those clients who do not speak English (Martinez, 2011); the shortages of O&M instructors (Mbulaheni et al., 2012), or the consequential decline in O&M skills of clients as a result of a paucity of resources and lack of instructors (Franck et al., 2011).

Because a majority of the articles on teaching O&M dealt with the implementation of high-tech tools, which was also found in the literature about individuals who are blind and visually impaired and have additional disabilities, this trend merited a deeper look. While the one study that examined a strategy of exploring a virtual environment and generalizing the knowledge gained in the virtual environment to the real environment with a modest learning curve found promising results (Lahav et al., 2012), the other studies that focused on technology found room for improvement. One study that used radar-like noises to create an audio representation of the environment was rated as highly usable by participants, but participants indicated the sounds emitted by the device needed to be optimized for better usability (Sánchez & Tadres, 2011). One study used a pocket computer to provide users with orientation information and information about the participant's location relative to other objects, but participants indicated they had better, low-tech ways of establishing orientation and needed more detailed information from the device before it could be useful (Sáenz & Sánchez, 2010). Participants in Ross and Kelly's (2009) study that focused on providing information to fill gaps about the environment also indicated that the information they received was not always helpful and that sometimes, they received too much information for their particular objectives. Although the device allowed participants to mute descriptions of the object in their environment that were not pertinent to their objective, the participants did not know whether the information would be helpful or not until after they heard it, so the mute feature was not particularly helpful (Ross & Kelly, 2009). Although most of the studies that focused on high-tech solutions for O&M training require further revision, they are all technologies that show promise if they are refined.

4.5. Multiple Disabilities, Including Deafblindness

In the category of O&M for individuals with multiple disabilities including deafblindness, 11 articles were identified and located. Four of these articles were research reports (Alaerts, Swinnen, & Wenderoth, 2011; Lancioni et al., 2009; Lancioni, O'Reilly, et al., 2010; Lancioni, Singh, et al., 2010), one presented the results of a preexperimental pilot study

(Bourquin, Hogan, & Sauerburger, 2010), one reviewed the literature (Parker, 2009), and five of the articles reported on current trends in practice (Bourquin & Moon, 2008; Erin, 2010; Feinberg, 2011; Griffin-Shirley & Pogrund, 2010; McKensie, 2010).

Of the four papers that reported the results of research, three used single-subject methodology (Lancioni et al., 2009; Lancioni, O'Reilly, et al., 2010; Lancioni, Singh, et al., 2010) and one used a quasiexperimental group design (Alaerts et al., 2011). Interestingly, Parker's (2009) literature review specifically examined single-subject studies that were published in peer-reviewed journals between the years of 1988 and 2007; thirteen studies were identified that met the author's criteria. All of the studies identified in Parker's review were by the same lead author of two single-subject studies identified in this chapter. In total, the author reviewed 860 articles, but did not report how many were excluded from the review because they employed a group design rather than a single-subject design. It stands to reason, however, that a majority of experimental studies that examine O&M with individuals with multiple disabilities including deafblindness would employ single-subject design due to the heterogeneous population of individuals who are visually impaired with multiple disabilities (Parker, 2009).

All four of the studies and the pilot studies that involved participants with multiple disabilities (including deafblindness) used adult participants (Alaerts et al., 2011; Bourquin et al., 2010; Lancioni et al., 2009; Lancioni, O'Reilly, et al., 2010; Lancioni, Singh, et al., 2010). The majority of participants were also congenitally blind. The one exception was the second study embedded in Lancioni, O'Reilly, et al. (2010). The only author that reported working with children rather than adults was the practice report by Feinberg (2011) that discussed a strategy for transitioning teenagers. All studies took place in rehabilitation centers or lab settings.

Some of the topics that were covered were the organic nature of what happens in the brain (Alaerts et al., 2011), the role of single-subject designs related to individuals who are deafblind (Parker, 2009), and soliciting aid when crossing streets (Bourquin et al., 2010). Three of the four practice reports focused on the inclusion of strategies when working with individuals with cortical visual impairment in O&M personnel preparation programs (Erin, 2010; Griffin-Shirley & Pogrund, 2010; McKensie, 2010), while one encouraged trying "out-of-the-box" strategies (Feinberg, 2011). The majority of the research tended to focus on the use of technology for wayfinding (Lancioni et al., 2009; Lancioni, O'Reilly, et al., 2010; Lancioni, Singh, et al., 2010). All three of these studies that focused on the use of technology for wayfinding used simple technology that provided auditory or auditory plus tactile cues. In all cases, the participants showed a significant improvement when using the technology, inconsistent with the research with participants who were blind or visually impaired without additional disabilities that showed fewer gains in O&M skills when using more high-tech tools.

5. DISCUSSION

There are two directions for future research: building on current research and filling gaps where there is currently little or no research. Both are necessary in any field. Building on current research is needed to generalize findings to new populations or situations and to replicate previous findings to build validity. Filling in gaps in research is necessary to build the theoretical base of a field, to address basic questions of what constitutes best practice from a quantitative foundation, and to assess new instructional practices and tools.

5.1. Building on Current Research

One of the major focus areas for individuals with visual impairments with and without additional disabilities was the use of technology for O&M. In the current literature, low-tech devices certainly have benefit for participants, but there is a consistent theme across topical areas to make use of newer technology. High-tech devices showed promise in teaching (e.g., Lahav et al., 2012; Ross & Kelly, 2009; Sáenz & Sánchez, 2010; Sánchez & Tadres, 2011), cane design (e.g., Bousbia-Salah et al., 2011), electronic travel aids (e.g., Havik et al., 2010; Hoefer, 2011; Kumar et al., 2013; Lahav & Mioduser, 2004, 2007; Lahav et al., 2012; Lancioni, Singh, et al., 2010; Penrod et al., 2010; Ponchillia, MacKenzie, et al., 2007; Ponchillia, Rak, et al., 2007; Roentgen et al., 2011, 2012; Ross & Kelly, 2009; Sáenz & Sánchez, 2010; Sánchez & Tadres, 2011; Simonnet & Vielledent, 2012; Wood, 2011), and accessibility (e.g., Hastings et al., 2011; Kim et al., 2012a; Ping et al., 2012; Schroeder et al., 2011; Wall Emerson et al., 2013). Because technology is being used in such a wide variety of areas and in such a wide variety of ways, much more research needs to be done in the area of technology use in O&M. Several researchers addressed how to teach the use of high-tech devices (e.g., Delgado Greenberg & Kuns, 2012; Phillips, 2011), an area of particular need for expansion. As the speed of technological development increases and the number of new devices and applications expands, effective approaches for how to assess the usefulness of devices and how to best teach consumers to use them are needed.

Another area that needs expansion is research on services for individuals with visual impairments and additional disabilities. In Parker's (2009) literature review, all of the studies were conducted by one primary researcher who also conducted the majority of studies identified by this chapter (Lancioni et al., 2009; Lancioni, O'Reilly, et al., 2010; Lancioni, Singh, et al., 2010). While this author has certainly made a sizable contribution to the field, replication by other researchers would strengthen the current body of research. Furthermore, their research only examined individuals who had lost vision adventitiously. The needs and abilities of individuals who have been blind since birth and the needs of individuals who lose vision later in life can be quite different, so both areas require research.

5.2. Filling Gaps in the Research

There were no research articles on children with visual impairments and additional disabilities that met our criteria for inclusion in this chapter. Therefore, research on O&M for children who are blind and visually impaired with additional disabilities including deafblindness is needed. In the field of O&M, research with adults is often reimagined and redesigned for application with children. In a similar vein, reliance on small sample sizes or single-subject research means that more replication of research needs to be conducted to generalize findings to other populations and other situations or to simply expand the theoretical and pedagogical usefulness findings.

After an initial search of the literature, it seemed that another area with a significant amount of work was focused on early childhood. However, a closer look at the identified articles whittled the list of appropriate articles down to one study that investigated the reaching responses of infants to auditory cues versus tactile cues (Ihsen, Troester, & Brambring, 2010). This study has implications for O&M but is not directly related. The only other publication that addressed early childhood was a practice report about incorporating music and rhythm to enhance O&M education (Sapp, 2011). The paucity of researchers examining early childhood represents an important gap in the literature, especially given the concern that children who are blind might suffer later in life if they do not develop appropriately in early childhood (Fraiberg, 1971).

5.3. Future Directions

From our synthesis of recent research in the field of O&M, it seems that using and adapting technology is a dominant trend. This burgeoning technology use is leading to accessibility in a greater variety of environments and access to a greater wealth of information. The increase in technological diversity and sophistication will tax instructors and consumers alike. Careful planning needs to be made regarding the usefulness of any given aid before limited resources of time and money are devoted to it. This means the field will need to develop a more comprehensive structure of assessment and planning for instruction. A greater diversity of potentially useful tools and strategies requires a mindful process for determining what tool or strategy will be most effective for a given person, in a given environment, for a given purpose. While continuing to develop new techniques and tools, the field of O&M will need to develop more valid comprehensive assessment tools and links between these tools and planning schema for instructors.

Just as the field needs to look inward to solidify its knowledge base and recommended practices, it should also branch out and form more solid partnerships with other fields. In 2011, AER's Division of O&M conducted a survey of its members and asked what the top five areas of interest for conference programming and webinars were. The areas of highest interest were deafblind instruction, working with different eye conditions or populations, accessibility, collaborating with other AER Divisions, presenting and manuscript writing, and play skills or recreation (Association for the Education, 2011). The findings from this survey seem to suggest that O&M practitioners are not most interested in particular topics, per se, but want to learn more about networking, dealing with diversity, and sharing ideas and strategies. This desire for new ideas and sharing may reflect a field that is primed to branch out and tackle new issues. To enhance environmental accessibility, O&M instructors should engage with local traffic engineers, sitting on local planning boards, and provide workshops on the accessibility needs of people who are blind. O&M instructors working in schools and rehabilitation agencies work on teams with other professionals to provide collaborative services for children and adult clients. The field needs to continue and expand these collaborations to build a more solid infrastructure for O&M. Developing stronger national certification processes, developing access to third-party Medicare billing, and becoming an integral part of the process of rehabilitation can only be achieved by enlisting the assistance of other fields (e.g., medicine, politics, occupational therapy, and industrial psychology).

The need to develop a more solid field and branch out to partner with other fields will only be effective, however, if there are a number of practitioners to provide adequate services. Development in all of the content areas discussed in this chapter and the need for the field to branch out and connect with other fields are limited by the capability to train an increasing number of O&M instructors and build capacity in the field. Increased capacity will allow the field of O&M to further branch out not only to partner with other fields but also to standardize services across countries and cultures and provide increased services in developing countries. While there are currently innovative projects under way to address those goals, the needs are so great that the field is not likely to address all of the needs without some sort of "sea change" within the field. Research, both to expand current findings to new situations and to fill in gaps in knowledge, will both strengthen and advance the field of O&M.

REFERENCES

Alaerts, K., Swinnen, S. P., & Wenderoth, N. (2011). Action perception in individuals with congenital blindness or deafness: How does the loss of a sensory modality from birth affect perception-induced motor facilitation? *Journal of Cognitive Neuroscience*, *23*(5), 1080–1087.

Anderson, D. (2010). *Orientation and mobility, reading, and math: Analysis of data for children with visual impairment from the special education elementary longitudinal study*. Unpublished doctoral dissertation, Kalamazoo, MI: Western Michigan University.

Ashmead, D., Grantham, W., Maloff, E., Hornsby, B., Nakamura, T., Davis, T., et al. (2012). Auditory perception of motor vehicle travel paths. *Human Factors*, *54*, 437–453.

Association for the Education and Rehabilitation of the Blind and Visually Impaired. (2011). O&M Survey Summary. Retrieved from http://oandm.aerbvi.org/documents/2011OMSurveySummaryFinal_000.pdf.

Ballemans, J., Kempen, G. I. J. M., & Zijlstra, G. A. R. (2011). Orientation and mobility training for partially-sighted older adults using an identification cane: A systematic review. *Clinical Rehabilitation*, *25*, 880–891.

Barlow, J. M., Bentzen, B. L., & Bond, T. (2005). Blind pedestrians and the changing technology and geometry of signalized intersections: Safety, orientation, and independence. *Journal of Visual Impairment & Blindness*, *99*, 587–598.

Barlow, J. M., & Franck, L. (2005). Crossroads: Modern interactive intersections and accessible pedestrian signals. *Journal of Visual Impairment & Blindness*, *99*, 599–610.

Barrella, K., Besden, C., Crow, N., Greenberg, M. D., Shrieves, G., Smith, K. A., et al. (2011). Striving to provide innovative orientation and mobility services in times of diminishing resources. *Journal of Visual Impairment & Blindness*, *105*, 587–590.

Bosbach, S. R. (1988). Precane mobility devices. *Journal of Visual Impairment & Blindness*, *82*, 338–339.

Bourquin, G., Hogan, S. H., & Sauerburger, D. J. (2010). Street crossing signs: Travelers who are deaf-blind obtaining assistance when pedestrians are not present. *AER Journal*, *3*, 139–145.

Bourquin, E., & Moon, J. (2008). Studies on obtaining assistance by travelers who are deaf-blind. *Journal of Visual Impairment & Blindness*, *102*(6), 352–361.

Bourquin, E., Wall Emerson, R., & Sauerburger, D. (2011). Conditions that influence drivers' yielding behavior while crossing at uncontrolled intersections. *Journal of Visual Impairment & Blindness, 105*, 760–769.

Bousbia-Salah, M., Bettayeb, M., & Larbi, A. (2011). A navigation aid for blind people. *Journal of Intelligent & Robotic Systems, 64*(3–4), 387–400.

Breslauer, B. E. (2010). The case for providing orientation and mobility instruction while using a dog guide. *Journal of Visual Impairment & Blindness, 104*, 5–6.

Brown, C., & Bour, B. (1986). Movement analysis and curriculum for visually impaired preschoolers. *A resource manual for the development and evaluation of special programs for exceptional students: Vol. V–K.* Tallahassee, FL: Florida Bureau of Education for Exceptional Students Department of Education.

Clarke, K. (1988). Barriers or enablers? Mobility devices for visually impaired multihandicapped infants and preschoolers. *Education of the Visually Handicapped, 20*, 115–132.

Clarke, K. L., Sainato, D. M., & Ward, M. E. (1994). Travel performance of preschoolers: The effects of mobility training with a long cane versus a precane. *Journal of Visual Impairment & Blindness, 88*, 19–30.

Delgado Greenberg, M., & Kuns, J. (2012). A free program for using and teaching an accessible electronic wayfinding device. *Journal of Visual Impairment & Blindness, 106*, 703–705.

Eichorn, J. R., & Vigaroso, H. R. (1967). Orientation and mobility for preschool blind children. *International Journal for the Education of the Blind, 17*, 48–50.

Erin, J. N. (2009). Traveling by touch: How useful are tactile maps? *Journal of Visual Impairment & Blindness, 103*, 265–266.

Erin, J. N. (2010). Developing the university curriculum to include CVI: A work in progress at the University of Arizona. *Journal of Visual Impairment & Blindness, 104*, 656–657.

Fazzi, D., & Naimy, B. J. (2010). Teaching orientation and mobility to school-age children. In W. R. Wiener, R. L. Welsh, & B. B. Blasch (Eds.), *Instructional strategies and practical applications: Vol. II. Foundations of orientation and mobility* (pp. 208–262). (3rd ed.). New York, NY: American Foundation for the Blind.

Feinberg, J. (2011). Becoming commuters: Teaching students traveling to work using public transportation. *Journal of Visual Impairment & Blindness, 105*, 262–264.

Ferrell, K. (1979). Orientation and mobility for preschool children: What we have and what we need. *Journal of Visual Impairment & Blindness, 73*, 147–150.

Foy, C. J., Kirchner, D., & Waple, L. (1991). The Connecticut precane. *Journal of Visual Impairment & Blindness, 85*, 85–86.

Fraiberg, S. (1971). Intervention in infancy: A program for blind infants. *Journal of the American Academy of Child Psychiatry, 10*, 381–405.

Franck, L., Haneline, R., & Farrugia, C. (2011). Canaries in the coal mine? Dog guide schools look at the current state of orientation and mobility. *Journal of Visual Impairment & Blindness, 105*(11), 741–742.

Garay-Vega, L., Hastings, A., Pollard, J., Zuschlag, M., & Stearns, M. (2010). *Quieter cars and the safety of blind pedestrians: Phase I (DOT HS 811 304).* Washington, DC: National Highway Traffic Safety Administration.

Geruschat, D. R., Fujiwara, K., & Wall Emerson, R. S. (2011). Traffic gap detection for pedestrians with low vision. *Optometry and Vision Science, 88*(2), 208–216.

Griffin-Shirley, N., & Pogrund, R. (2010). Inclusion of CVI in Texas Tech University's personnel preparation program. *Journal of Visual Impairment & Blindness, 104*, 660–661.

Gross, F., Lyon, C., Persaud, B., & Srinivasan, R. (2013). Safety effectiveness of converting signalized intersections to roundabouts. *Accident Analysis and Prevention, 50*, 234–241.

Guth, D., Long, R., Wall Emerson, R., Ashmead, D., & Ponchillia, P. (2013). Blind and sighted pedestrians' judgments of gaps in traffic at a single lane roundabout: Effects of traffic volume and pedestrian location. *Human Factors, 55*(3), 632–642.

Hanna, R. (2009). *Incidence of pedestrian and bicyclist crashes by hybrid electric passenger vehicles (DOT HS 811 204)*. Washington, DC: National Highway Traffic Safety Administration.

Hapeman, L. B. (1967). Developmental concepts of blind children between the ages of three and six as they relate to orientation and mobility. *International Journal for the Education of the Blind, 17*, 41–48.

Hassan, S. E., & Massof, R. W. (2012). Measurements of street-crossing decision-making in pedestrians with low vision. *Accident Analysis and Prevention, 49*, 410–418.

Hastings, A., Pollard, J. K., Garay-Vega, L., Stearns, M. D., & Guthy, C. (2011). *Quieter cars and the safety of blind pedestrians, Phase 2: Development of potential specifications for vehicle countermeasure sounds—Final report (DOT HS 811 496)*. Washington, DC: National Highway Traffic Safety Administration.

Hatlen, P. (2000). Historical perspectives. In M. C. Holbrook, & A. J. Koenig (Eds.), *Foundations of education volume 1: History and theory of teaching children and youths with visual impairments* (pp. 1–54) (2nd ed.). New York, NY: American Foundation for the Blind.

Havik, E. M., Kooijman, A. C., & Steyvers, F. J. J. M. (2011). The effectiveness of verbal information provided by electronic travel aids for visually impaired persons. *Journal of Visual Impairment & Blindness, 105*, 624–637.

Havik, E. M., Steyvers, F. J. J. M., van der Velde, H., Pinkster, J. C., & Kooijman, A. C. (2010). Design and evaluation of a protocol to assess electronic travel aids for persons who are visually impaired. *Journal of Visual Impairment & Blindness, 104*, 84–94.

Hill, E. W. (1988). *Preschool orientation and mobility project for visually impaired children (final report)*. Nashville, TN: Peabody College, Vanderbilt University.

Hill, E. W., Dodson-Burke, B., & Smith, B. (1989). Orientation and mobility for infants who are visually impaired. *RE:view, 21*, 57–60.

Hill, E. W., Rosen, S., Correa, V. I., & Langley, M. B. (1984). Preschool orientation and mobility: An expanded definition. *Education of the Visually Handicapped, 16*, 58–72.

Hoefer, S. (2011). Developing an open source sonar navigation device. *International Journal of Orientation and Mobility, 4*, 95–98.

Hong, S., Cho, K., & Ko, B. (2013). Investigation of probability of pedestrian crash based on auditory recognition distance due to a quiet vehicle in motor mode. *International Journal of Automotive Technology, 14*(3), 441–448.

Huang, K., Leung, C., & Wang, H. (2010). Effects of cane length and diameter and judgment type on the constant error ratio for estimated height in blindfolded, visually impaired, and sighted participants. *Perceptual and Motor Skills, 110*(2), 593–602.

Ihsen, E., Troester, H., & Brambring, M. (2010). The role of sound in encouraging infants with congenital blindness to reach for objects. *Journal of Visual Impairment & Blindness, 104*(8), 478–488.

Kim, D. S., & Wall Emerson, R. (2012). Effect of cane length on drop-off detection performance. *Journal of Visual Impairment & Blindness, 106*(1), 31–35.

Kim, D. S., Wall Emerson, R., & Curtis, A. (2009). Drop-off detection with the long cane: Effects of different cane techniques on performance. *Journal of Visual Impairment & Blindness, 103*, 519–530.

Kim, D. S., Wall Emerson, R., & Curtis, A. (2010a). Analysis of user characteristics related to drop-off detection with long cane. *Journal of Rehabilitation Research and Development, 47*(3), 233–242.

Kim, D. S., Wall Emerson, R. S., & Curtis, A. (2010b). Ergonomic factors related to drop-off detection with the long cane: Effects of cane tips and techniques. *Human Factors, 52*, 456–465.

Kim, D. S., Wall Emerson, R., & Curtis, A. (2010c). Interaction effects of the amount of practice, preferred cane technique, and type of cane technique used on drop-off detection performance. *Journal of Visual Impairment & Blindness, 104*(8), 453–463.

Kim, D. S., Wall Emerson, R., Naghshineh, K., Pliskow, J., & Myers, K. (2012a). Impact of adding an artificially generated alert sound to hybrid electric vehicles on their detectability by pedestrians who are blind. *Journal of Rehabilitation Research and Development, 49*(3), 381–394.

Kim, D. S., Wall Emerson, R., Naghshineh, K., Pliskow, J., & Myers, K. (2012b). Vehicle surge detection and pathway discrimination by pedestrians who are blind: Effect of adding an alert sound to hybrid electric vehicles on performance. *British Journal of Visual Impairment, 30*(2), 61–78.

Kronick, M. K. (1987). Children and canes: An adaptive approach. *Journal of Visual Impairment & Blindness, 81*, 61–62.

Kumar, A., Patra, R., Mahadevappa, M., Mukhopadhyay, J., & Majumdar, A. K. (2013). An embedded system for aiding navigation of visually impaired persons. *Current Science (Bangalore), 104*(3), 302–306.

La Grow, S. J., Ihrke, E., Ponchillia, P. E., Sullins, C. D., Owiti, S. A., & Lewis, L. (2009). User perceptions of accessible GPS as a wayfinding tool. *AER Journal, 2*, 111–120.

Lahav, O., & Mioduser, D. (2004). Exploration of unknown spaces by people who are blind using a multi-sensory virtual environment. *Journal of Special Education Technology, 19*(3), 15–23.

Lahav, O., & Mioduser, D. (2007). Haptic-feedback support for cognitive mapping of unknown spaces by people who are blind. *International Journal of Human-Computer Studies, 66*, 23–35.

Lahav, O., Schloerb, D. W., & Srinivasan, M. A. (2012). Newly blind persons using virtual environment system in a traditional orientation and mobility rehabilitation program: A case study. *Disability and Rehabilitation. Assistive Technology*, 7(5), 420–435.

Lancioni, G. E., O'Reilly, M. F., Singh, N. N., Sigafoos, J., Campodonico, F., & Oliva, D. (2009). A wheelchair user with visual and intellectual disabilities managing simple orientation technology for indoor travel. *Journal of Visual Impairment & Blindness, 103*, 308–313.

Lancioni, G. E., O'Reilly, M. F., Singh, N. N., Sigafoos, J., Campodonico, F., & Oliva, D. (2010). Two persons with multiple disabilities use orientation technology with auditory cues to manage simple indoor traveling. *Research in Developmental Disabilities, 31*, 397–402.

Lancioni, G. E., Singh, N. N., O'Reilly, M. F., Sigafoos, J., Alberti, G., Scigliuzzo, F., et al. (2010). Persons with multiple disabilities use orientation technology to find room entrances during indoor traveling. *Research in Developmental Disabilities: A Multidisciplinary Journal, 31*(6), 1577–1584.

Long, R. G., Guth, D. A., Ashmead, D. H., Wall Emerson, R., & Ponchillia, P. E. (2005). Modern roundabouts: Access by pedestrians who are blind. *Journal of Visual Impairment & Blindness, 99*(10), 611–621.

Mancil, R., Manuel, S., Sifferman, E., & Blasch, B. (1998). Results of a study to develop an alternative cane. *RE:view, 30*, 130–138.

Martinez, D. (2011). New practitioner identifies the need for more multi-lingual orientation and mobility instructional materials. *Journal of Visual Impairment & Blindness, 105*, 583–584.

Mbulaheni, M. O., Dzapasi, A., & Sabeya, P. (2012). Orientation and mobility services for persons with visual impairments: South African orientation and mobility practitioners' eye view. *Journal of Visual Impairment & Blindness, 106*(11), 750–755.

McKensie, A. R. (2010). Personnel preparation for training professionals to work with individuals with CVI at Florida State University. *Journal of Visual Impairment & Blindness, 104*, 655–656.

New Jersey Council of the Blind. (n.d.). The history of the white cane. Retrieved from http://www.njcounciloftheblind.org/brochures/history_of_white_cane.htm.

O'Mara, B. (1989). *Pathways to independence: Orientation and mobility skills for your infant and toddler.* New York, NY: The Lighthouse.
O'Mea, M. L. (2013). Implementing applied behavior analysis for effective orientation and mobility instruction of students with multiple disabilities. *Journal of Visual Impairment & Blindness, 107*(1), 65–70.
Parker, A. T. (2009). Orientation and mobility with persons who are deaf-blind: An initial examination of single-subject design research. *Journal of Visual Impairment & Blindness, 103*, 372–377.
Pedestrian Safety Enhancement Act of 2010. Pub. L. 111–373. Available at http://www.gpo.gov/fdsys/pkg/PLAW-111publ373/content-detail.html.
Penrod, W. M. (2012). A seamless approach to transitioning cane skills from the diagonal to the two-point touch technique. *Journal of Visual Impairment & Blindness, 106*, 235–239.
Penrod, W. M., Simmons, T. J., Bauder, D. K., & Brostek Lee, D. (2009). A comparison of selected secondary electronic travel aids with a primary mobility system. *International Journal of Orientation and Mobility, 2*, 27–41.
Penrod, W. M., Simmons, T. J., Bauder, D. K., Brostek Lee, D., Haneline, R., Abraham, H., et al. (2010). The effectiveness of master trainer courses to teach electronic mobility aids. *AER Journal, 3*, 20–25.
Phillips, C. L. (2011). Getting from here to there and knowing where: Teaching global positioning systems to students with visual impairments. *Journal of Visual Impairment & Blindness, 105*, 675–680.
Ping, K. P., Diew, W. Y., Menow, A. P. G., & Palanusamy, C. (2012). Mobility of visually impaired pedestrians crossing behavior and assistive design/technologies at signalized pedestrian crossings. *Institute of Transportation Engineers. ITE Journal, 82*, 33–37.
Pogrund, R. L., & Rosen, S. J. (1989). The preschool blind child *can* be a cane user. *Journal of Visual Impairment & Blindness, 83*, 431–439.
Ponchillia, P. E., MacKenzie, N., Long, R. G., Denton-Smith, P., Hicks, T. L., & Miley, P. (2007). Finding a target with an accessible global positioning system. *Journal of Visual Impairment & Blindness, 101*, 479–488.
Ponchillia, P. E., Rak, E. C., Freeland, A. L., & LaGrow, S. J. (2007). Accessible GPS: Reorientation and target location among users with visual impairments. *Journal of Visual Impairment & Blindness, 101*, 389–401.
Robart, R. L., & Rosenblum, L. D. (2009). Are hybrid cars too quiet? *Journal of the Acoustical Society of America, 125*, 2744.
Rodgers, M. D., & Wall Emerson, R. (2005a). Human factor analysis of long cane design: Weight and length. *Journal of Visual Impairment & Blindness, 99*, 622–632.
Rodgers, M. D., & Wall Emerson, R. (2005b). Materials testing in long cane design: Sensitivity, flexibility, and transmission of vibration. *Journal of Visual Impairment & Blindness, 99*(11), 696–706.
Roentgen, U. R., Gelderblom, G. J., & de Witte, L. P. (2011). Users' evaluations of four electronic travel aids aimed at navigation for persons who are visually impaired. *Journal of Visual Impairment & Blindness, 105*, 612–623.
Roentgen, U. R., Gelderblom, G. J., & de Witte, L. P. (2012). The development of an indoor mobility course for the evaluation of electronic mobility aids for persons who are visually impaired. *Assistive Technology, 24*, 143–154.
Roentgen, U. R., Gelderblom, G. J., Soede, M., & de Witte, L. P. (2008). Inventory of electronic mobility aids for persons with visual impairments: A literature review. *Journal of Visual Impairment & Blindness, 102*, 702–724.
Ross, D. A., & Kelly, G. W. (2009). Filling the gaps for indoor wayfinding. *Journal of Visual Impairment & Blindness, 103*, 229–234.
Sáenz, M., & Sánchez, J. (2010). Indoor orientation and mobility for learners who are blind. *Annual Review of CyberTherapy and Telemedicine, 8*, 131–134.

Sanchez, J., & Saenz, M. (2010). Metro navigation for the blind. *Computers & Education, 55*(3), 970–981.
Sánchez, J., & Tadres, A. (2011). Augmented reality application for the navigation of people who are blind. *The International Journal on Disability and Human Development, 10*(1), 75–79.
Sapp, W. (2011). Somebody's jumping on the floor: Incorporating music into orientation and mobility for preschoolers with visual impairments. *Journal of Visual Impairment & Blindness, 105*, 715–719.
Sauerbuger, D. (2006). Instructional strategies for teaching judgment in detecting gaps for crossing streets with no traffic controls. *RE:view, 37*, 177–188.
Sauerburger, D. (2005). Street crossings: Analyzing risks, developing strategies, and making decisions. *Journal of Visual Impairment & Blindness, 99*, 659–663.
Sauerburger, D., & Bourquin, E. (2010). Teaching the use of a long cane step by step: Suggestions for progressive, methodical instruction. *Journal of Visual Impairment & Blindness, 104*, 203–214.
Schroeder, B., Hughes, R., Rouphail, N., Cunningham, C., Salamati, K., Long, R., et al. (2011). *NCHRP Report 674: Crossing solutions at roundabouts and channelized turn lanes for pedestrians with vision disabilities.* Washington, DC: Transportation Research Board.
Scott, B. (2010). Early long cane use: A case study. *AER Journal, 3*, 26–29.
Scott, A. C., Barlow, J. M., Guth, D. A., Bentzen, B. L., Cunningham, C. M., & Long, R. (2011a). Nonvisual cues for aligning to cross streets. *Journal of Visual Impairment & Blindness, 105*, 648–661.
Scott, A. C., Barlow, J. M., Guth, D. A., Bentzen, B. L., Cunningham, C. M., & Long, R. (2011b). Walking between the lines: Nonvisual cues for maintaining headings during street crossings. *Journal of Visual Impairment & Blindness, 105*, 662–674.
Simonnet, M., & Vielledent, S. (2012). Accuracy and coordination of spatial frames of reference during the exploration of virtual maps: Interest for orientation and mobility of blind people? *Advances in Human-Computer Interaction, 2012*, 1–14. http://dx.doi.org/10.1155/2012/835246, Article ID 835246.
Skellenger, A. C., & Sapp, W. K. (2010). Teaching orientation and mobility for the early childhood years. In W. R. Wiener, R. L. Welsh, & B. B. Blasch (Eds.), *Instructional strategies and practical applications: Vol. II. Foundations of orientation and mobility* (pp. 163–207). (3rd ed.). New York, NY: American Foundation for the Blind.
Smith, D. L., & Penrod, W. M. (2010). Adaptive technology for orientation and mobility. In W. R. Wiener, R. L. Welsh, & B. B. Blasch (Eds.), *History and theory: Vol. I. Foundations of orientation and mobility* (pp. 241–276). (3rd ed.). New York, NY: American Foundation for the Blind.
Strickfaden, M., & Devlieger, P. (2011). The Brussels metro: Accessibility through collaboration. *Journal of Visual Impairment & Blindness, 105*(10), 638–647.
Wall Emerson, R., Kim, D. S., Naghshineh, K., Pliskow, J., & Myers, K. (2011). Examining the impact of "quiet" vehicles on the performance of orientation and mobility tasks by blind pedestrians. *Journal of Visual Impairment and Blindness, 105*(11), 745–759.
Wall Emerson, R., Kim, D. S., Naghshineh, K., Pliskow, J., & Myers, K. (2013). Detection of "quiet" vehicles by blind pedestrians. *Journal of Transportation Engineering, 139*(1), 50–56.
Wall Emerson, R., Naghshineh, K., Hapeman, J., & Weiner, W. (2011). A pilot study of pedestrians with visual impairments detecting traffic gaps and surges containing hybrid vehicles. *Transportation Research Part F: Traffic Psychology and Behavior, 14*(2), 117–127.
Wiener, W. R., & Sifferman, E. (2000). A demographic study of certified orientation and mobility specialists. *RE:view, 32*, 39–45.
Wiener, W. R., & Sifferman, E. (2010). The history and progression of the profession of orientation and mobility. In W. R. Wiener, R. L. Welsh, & B. B. Blasch (Eds.), *History*

and theory: Vol. I. Foundations of orientation and mobility (pp. 486–532). (3rd ed.). New York, NY: American Foundation for the Blind.

Wiener, W. R., Welsh, R. L., & Blasch, B. B. (2010). Introduction. In W. R. Wiener, R. L. Welsh, & B. B. Blasch (Eds.), *History and theory: Vol. I. Foundations of orientation and mobility* (p. xv). (3rd ed.). New York, NY: American Foundation for the Blind.

Willins, K. (2012). The spring loaded "Willins" cane. *International Journal of Orientation and Mobility, 5*, 40–43.

Wong, M. E., & Tan, S. S. K. (2012). Teaching the benefits of smart phone technology to blind consumers: Exploring the potential of the iPhone. *Journal of Visual Impairment & Blindness, 106*(10), 646–650.

Wood, M. (2011). Trekker Breeze 2.0: Trialled by clients. *International Journal of Orientation and Mobility, 4*, 81–86.

Wright, T., Harris, B., & Sticken, E. (2010). A best-evidence synthesis of research on orientation and mobility involving tactile maps and models. *Journal of Visual Impairment & Blindness, 104*, 95–106.

Wright, T. S., & Wolery, M. (2014). Evaluating the effectiveness of roadside instruction in teaching youth with visual impairments street crossings. *The Journal of Special Education, 48*, 46–58.

Zijlstra, G. A. R., Ballemans, J., & Kempen, G. I. J. M. (2013). Orientation and mobility training for adults with low vision: A new standardized approach. *Clinical Rehabilitation, 27*(1), 3–18.

INDEX

Note: Page numbers followed by "*f*" indicate figures and "*t*" indicate tables.

P

Q

R

S

CONTENTS OF PREVIOUS VOLUMES

VOLUME 1

VOLUME 2

VOLUME 3

VOLUME 4

VOLUME 5

VOLUME 6

VOLUME 7

VOLUME 8

VOLUME 9

VOLUME 10

VOLUME 11

VOLUME 12

VOLUME 13

VOLUME 14

VOLUME 15

VOLUME 16

VOLUME 17

VOLUME 18

VOLUME 19

VOLUME 20

VOLUME 21

VOLUME 22

VOLUME 23

VOLUME 24

VOLUME 25

VOLUME 26

VOLUME 27

VOLUME 28

VOLUME 29

VOLUME 30

VOLUME 31

VOLUME 32

VOLUME 33

VOLUME 34

VOLUME 35

VOLUME 36

VOLUME 37

VOLUME 38

VOLUME 39

VOLUME 40

VOLUME 41

VOLUME 42

Kolibri 360 GmbH
Ferdinand-Jühlke-Straße 7
99095 Erfurt, Deutschland
produktsicherheit@kolibri360.de